The Biology of Lakes and Ponds

BIOLOGY OF HABITATS

Series editors: M. Crawley, C. Little,
T. R. E. Southwood, and S. Ulfstrand

The intention is to publish attractive texts giving an integrated overview of the design, physiology, ecology, and behaviour of the organisms in given habitats. Each book will provide information about the habitat and the types of organisms present, on practical aspects of working within the habitats and the sorts of studies which are possible, and will include a discussion of biodiversity and conservation needs. The series is intended for naturalists, students studying biological or environmental sciences, those beginning independent research, and biologists embarking on research in a new habitat.

The Biology of Lakes and Ponds

Christer Brönmark

Department of Ecology
University of Lund, Sweden

and

Lars-Anders Hansson

Department of Ecology
University of Lund, Sweden

OXFORD
UNIVERSITY PRESS

OXFORD

UNIVERSITY PRESS

Great Clarendon Street, Oxford OX2 6DP

Oxford University Press is a department of the University of Oxford.
It furthers the University's objective of excellence in research, scholarship,
and education by publishing worldwide in

Oxford New York

Auckland Cape Town Dar es Salaam Hong Kong Karachi
Kuala Lumpur Madrid Melbourne Mexico City Nairobi
New Delhi Shanghai Taipei Toronto

With offices in

Argentina Austria Brazil Chile Czech Republic France Greece
Guatemala Hungary Italy Japan Poland Portugal Singapore
South Korea Switzerland Thailand Turkey Ukraine Vietnam

Oxford is a registered trade mark of Oxford University Press
in the UK and in certain other countries

Published in the United States
by Oxford University Press Inc., New York

First editon published 1998

Reprinted 1999, 2000, 2001, twice 2002, 2003, 2004

Second edition published 2005

Reprinted 2005 (with corrections), 2006

A catalogue record for this title is available from the British Library

Library of Congress Cataloging-in-Publication Data
Brönmark, Christer.
    The biology of lakes and ponds / Christer Brönmark, Lars-Anders Hansson.—2nd ed.
        p. cm.
    ISBN 0–19–851612–6 (alk. paper)—ISBN 0–19–851613–4 (alk. paper) 1.   Limnology.
2.   Lake ecology. 3.   Pond ecology. I.   Hansson, Lars-Anders. II.   Title.
    QH96.B724 2005
    577.63—dc22   2004025884

Typeset by Newgen Imaging Systems (P) Ltd., Chennai, India
Printed in Great Britain
on acid-free paper by Biddles, King's Lynn.

ISBN 0 19 851612 6 (Hbk)   978 0 19 851612 5
ISBN 0 19 851613 4 (Pbk)   978 0 19 851613 2

10 9 8 7 6 5 4 3

Foreword

Lakes and ponds cover only a very small portion of the earth's surface, yet their study has played a disproportionately large role in the development of ecological theory. From broad ecosystem-level concepts such as trophic structure and trophic dynamics (Elton 1927, Lindeman 1942, Hutchinson 1948), the trophic cascade (Zaret and Paine 1973, Carpenter et al. 1985), and ecological stoichiometry (Tilman 1982, Sterner and Elser 2002), to the impacts of competition and predation in community structure (Hutchinson 1961, Hrbáek 1961, Brooks and Dodson 1965), to key features of species adaptations such as induced morphological and behavioral defenses (Gilbert 1967, Krueger and Dodson 1981, Dodson 1988), fundamental discoveries have flowed from the study of these relatively small lenses of freshwater resting on our planet's crust.

One reason for this is that lakes and ponds are superb natural laboratories for understanding how the ecological world functions. Like plants and animals on ocean islands, the populations and communities of organisms in lakes are more isolated from exchange with nearby habitats than are their terrestrial counterparts. The microbes, algae, plants and animals that live submersed beneath the water's surface are nearly all restricted to their liquid environment, unable to cross the land that surrounds them. Thus confined, their interactions are often intense, easily discerned, and complex in fascinating ways. The diversity of genetic and species assemblages to be found among lakes and ponds, even within close proximity to each other, has provided exceptional opportunities for dissecting the forces that produce variation in nature.

The thought that lakes and ponds are isolated from the surrounding landscape goes back at least to Forbes' classic 1887 paper "The lake as a microcosm" (reprinted 1925). This independence, of course, should not be overstated (Likens 1984, Wetzel 1990): for example, the population genetic structure of even obligately aquatic organisms suggests occasional exchange of individuals, and there are certainly animal taxa that regularly move from lake to lake or between water and land. In addition, there are substantial subsidies of lake primary and secondary production by organic and inorganic chemicals flowing from the watershed (Pace et al. 2004). Indeed, the sediments of lakes store the historical record of a wide variety of ecological and evolutionary events that have taken place not only inside the confines of the lake proper (Kilham et al. 1986, Hairston et al. 1999), but also within its watershed and beyond (Davis ans Shaw 2001). Yet, as Forbes (1887) noted so long ago: "One finds in a single body of water a far more complete and independent equilibrium of organic life and activity than on any equal body of land." The idea of a tight interdependence among the organisms resident in lakes that he described has remained both a significant paradigm and a useful foil for evolutionary biologists, ecologists and limnologists studying the adaptations, abundance, species composition and trophic interactions of lake ecosystems.

A second reason underlying the central role of lakes in the formation of basic ecological understanding lies in the nature of water itself: it is a solvent for organic and inorganic molecules, a trap for heat, a filter for light, and a dense and viscous medium in which fluid movements large and small play a critical role. As a result of this tight coupling between the physical, chemical and biological world in lakes, limnologists were among the first to take a whole-ecosystem approach to ecology. When August Thienemann (1925) tied lake basin shape, nutrient and oxygen concentrations, and organism abundance together in a single conceptual framework, he summarized the accumulated knowledge of a field that was already some four decades old (Forbes 1887, Forel 1892). This linkage between geological, chemical, physical and biological processes remains a hallmark of limnology.

Christer Brönmark and Lars-Anders Hansson are ideal scientists to summarize the diversity of interactions among freshwater organisms and the aquatic context in which these links take place. Their research has been among the most illuminating in showing both the important and fascinating ways by which organisms from microscopic primary producers to large predatory fish detect and respond to each other, and how these interactions play out in the broader community and ecosystem context. The first edition of their book received well-deserved praise as a readable text for an undergraduate course and a useful summary of the ecological aspects of limnology for graduate students and instructors (Knight 2000, Kilham 2000). In this second edition, Brönmark and Hansson expand their coverage of population and community processes, link these to whole-ecosystem dynamics, and address a comment of the first edition by strengthening their coverage of the physical and chemical context in which the intra- and interspecific interactions that they expertly summarize take place. Each of these additions enhances the value of this book as an introduction first to the ecology of lakes, and second to lakes as model systems for studying general ecological processes. In addition, Brönmark and Hansson have expanded their final chapter on the challenges that human activities, intentional and unintended, have on lake and pond ecosystems. It is the basic ecological and limnological knowledge captured so well by the book you have in your hands that makes possible informed decisions about how to manage lakes and ponds.

Nelson Hairston Jr.
Department of Ecology and Evolutionary Biology,
Cornell University

Literature cited

Brooks, J. L. and Dodson, S. I. (1965). Predation, body size, and composition of plankton. *Science*, **150**, 28–35.

Carpenter, S. R., Kitchell, J. F., and Hodgson, J. R. (1985). Cascading trophic interactions and lake productivity: Fish predation and herbivory can regulate lake ecosystems. *Bioscience*, **35**, 634–9.

Davis, M. B. and R. G. Shaw. 2001. Range shifts and adaptive responses to Quarternary climate change. *Science*, **292**, 673–9.

Dodson, S. I. (1988) The ecological role of chemical stimuli for zooplankton: predator-avoidance behavior in Daphnia. *Limnology and Oceanography*, 33, 1431–9.

Elton, C. (1927). Animal ecology. MacMillan Co., New York

Forbes, S. (1925). The lake as a microcosm. *Bulletin of the Illinois Natural History Survey*, **15**, 537–50. (Originally published 1887)

Forel F.-A. (1892). Le Léman: monographie limnologique. Tome I. Géographie, hydrographie, géologie, climatologie, hydrologie, Lausanne, F. Rouge, 543 pp.

Gilbert, J. J. (1967). *Asplanchna* and postero-lateral spine production in *Brachionus calyciflorus*. *Archiv für Hydrobiologie*, **64**, 1-62.

Hairston, N. G., Jr., W. Lampert, C. E. Cáceres, C. L. Holtmeier, L. J. Weider, U. Gaedke, J. M. Fischer, J. A. Fox, and D. M. Post. 1999. Rapid evolution revealed by dormant eggs. *Nature*, **401**, 446.

Hrbácek, J. (1962). Species composition and the amount of zooplankton in relation to the fish stock. *Rozpr. Ceskoslovenské Akademie Ved Rada Mathematickch a Prírodních Ved*, *72*, 1–114.

Hutchinson, G. E. (1948). Circular causal systems in ecology. *Annals of the New York Academy of Sciences*, **50**, 221–46.

Hutchinson, G. E. (1961). The paradox of the plankton. *American Naturalist*, **95**, 137–45.

Likens, G. E. (1984). Beyond the shoreline: A watershed-ecosystem approach. *Internationale Vereinigung für theoretische and angewandte Limnologie*, **22**, 1-22.

Kilham, P., Kilham, S. S. and Hecky, R. E. (1986). Hypothesized resource relationships among African planktonic diatoms. *Limnology and Oceanography*, **31**, 1169–81.

Kilham, S. S. (2000). Review of "The Biology of Lakes and Ponds" by Brönmark and Hansson. *Limnology and Oceanography*, **45**, 752.

Knight, T. (2000). Review of "The Biology of Lakes and Ponds" by Brönmark and Hansson. *Ecology*, **81**, 286–7.

Krueger, D. A. and Dodson, S. I. (1981). Embryological induction and predation ecology in *Daphnia pulex*. *Limnology and Oceanography*, **26**, 219–23.

Lindeman, R. L. (1942). The trophic-dynamic aspect of ecology. *Ecology*, **23**, 399–418.

Pace, M. L., J. J. Cole, S. R. Carpenter, J. F. Kitchell, J. R Hodgson, M C. Van de Bogert, D. L. Bade, E. S. Kritzberg, and D. Bastviken. (2004). Whole-lake carbon-13 additoins reveal terrestrial support of aquatic food webs. *Nature*, **427**, 240–3.

Sterner, R. W., and Elser. J. J. (2002). Ecological stoichiometry: The biology of elements from molecules to the biosphere. Princeton Univ. Press. Princeton.

Thienemann, A. (1925) Die Binnengewässer Mitteleuropas. Ein limnologische Einführung. *Die Binnengewässer*, **1**, 225 pp.

Tilman, D. (1982). Resource competition and community structure. *Monographs in Population biology*. Princeton University Press, NJ.

Wetzel, R. G. (1990). Land-water interfaces: Metabolic and limnological regulators. *Internationale Vereinigung für theoretische and angewandte Limnologie*, **24**, 6–24.

Zaret, T. M. and Paine, R. T. (1973). Species introduction in a tropical lake. *Science*, **182**, 449–55.

Preface

Do you remember the fascination with the incredible diversity of life forms you found when you pulled a sweep-net through the nearshore plants of a small pond? Or the feeling you had the first time you caught a fish in your childhood pond or lake? And how the frog eggs you collected and kept in the small tank one day had hatched into tadpoles? And how these small, black 'lawnmowers' were feeding on the algal mat and then one day they had metamorphosed into small frogs? Or, remember the joy playing along the shore or in the shallow waters of a lake during endless summers; a joy mixed with disgust of the muddy bottom sediments and the creepy feeling of water plants winding up your legs, or the scary belief that below the surface awaits an armada of leeches and other yet unknown creatures prepared for attack! Indeed, our feelings for lakes and ponds are individual and contradictory, and thus, our relations with freshwater systems are coloured by a mixture of joy, mysticism, and even fright, but also by the knowledge that we to a large extent actually consist of water and that we therefore are a part of the endless merry-go-round of water through rain, landscapes, lakes, and organisms—the *hydrological cycle*. If you have ever touched upon such feelings, you have already taken a first step into understanding the intriguing world of freshwater systems! Man has always been fascinated by, but also dependent on, freshwater plants and animals. Freshwater organisms provide an important food source and freshwater in itself is of course an essential resource for human life. Thus, it is of crucial importance to study and understand the dynamics of freshwater systems and, consequently, a specific branch of science has been devoted to this, namely, *limnology* (derived from the Greek word for 'lake', *limnos*).

Traditional limnology has used a holistic, ecosystem approach to the study of freshwater systems, focusing on physicochemical processes at the lake level (lake metabolism, nutrient cycles, lake classification). However, there has been a conflict among scientists whether abiotic *or* biotic processes are most important in shaping lake ecosystems. For a long time it was considered that abiotic factors, such as temperature and nutrients, were the major determinants of structure and dynamics in lakes and ponds. Much less emphasis had been put on the importance of interactions between organisms, such as competition, predation, and complex interactions in food chains. But in recent years there has been an increasing interest in applying modern evolutionary and ecological theory to freshwater systems. This has resulted in a large number of studies on physiological, morphological, and behavioural adaptations among organisms to abiotic and biotic factors and how interactions between biotic processes and abiotic constraints determine the dynamics of freshwater systems.

Throughout the world there is a tremendous variation in the size and permanence of waterbodies. They range in size from small puddles to the huge Great Lakes of North America and Lake Baikal in Russia. Some lakes are millions of years old, whereas at the other extreme ephemeral waterbodies are water-filled only during part of the year or perhaps even only during especially wet years. We also find standing waterbodies in all climatic zones, from Arctic regions where lakes are ice-covered during most of the year, to the tropical regions that experience high temperatures year-round. Of course, this large variation in spatial and temporal scale as well as in climatic conditions creates a tremendous difference in the environmental conditions experienced by the organisms that inhabit these systems. Thus, organisms from different systems are expected to show a substantial diversity of adaptations to cope with their environment and to be specialized to a specific set of abiotic and biotic conditions. However, although organisms from different freshwater systems differ greatly, they have surprisingly many features of their ecology in common.

In this book we will focus on these general patterns in adaptations and processes found in freshwater systems. The emphasis is on lakes and ponds from temperate systems in Europe and North America, although we also have included examples from Antarctic, Arctic, and tropical regions. This is, of course, due to our own personal bias, but also partly a result of the fact that most studies have been performed in northern temperate regions. With respect to organisms, we use examples from animals, plants, and microbes wherever they are suitable to express or explain a certain process or principle. We have consciously given more emphasis to both larger (fish) and smaller organisms (ciliates, heterotrophic flagellates, and bacteria) than is the tradition in limnology textbooks. This expansion of the size spectrum of organisms is because recent research has pointed out their importance for the structure and dynamics of freshwater systems.

In addition to focusing on general patterns and adaptations, we want to introduce the most important features of both aquatic ecology and traditional limnology. This has been a major challenge as we only had a limited number of pages to work with. We soon discovered that including all the themes that we originally planned was impossible. Instead, we have chosen to identify the niche between traditional limnology and modern evolutionary ecology. Therefore, the book covers far from all aspects of traditional limnology. The focus is instead on adaptations among organisms to cope with abiotic factors and the importance of biotic processes and interactions in food webs. Thus, we have presented our own, highly subjective, distillation and blend of what we think is interesting and important to know for an aquatic ecologist at the beginning of his or her career. We have written the book mainly for the use of undergraduate students taking aquatic ecology/limnology courses but also for more advanced students who want to obtain an insight into the field of aquatic ecology. However, it is also our hope that it will be of interest to field naturalists and for those who enjoy strolling around lakes and ponds.

The first edition of this book has now been used as textbook in aquatic ecology and limnology courses for several years and we have received a lot of constructive comments and suggestions for changes from friends and colleagues. Accordingly, we have done some major changes in this 2nd revision. First, we have expanded the chapter "*The Abiotic Frame*" in order to strengthen the discussion of the effects of physical and chemical factors on the biology of lakes and ponds. Another major alteration in this edition is the expansion of chapter 6 ("*Biodiversity and Environmental threats*") which now focus more on the effects of environmental threats on the biodiversity of organisms. Finally, since the scientific frontiers move rapidly forward, all chapters in the 2nd edition have been upgraded with respect to new research findings and concepts. These changes have resulted in an increase in volume of about 25%, which we hope have made the book even more useful, allowing it to continue serving as inspiration for you who study, or simply are interested in, the Biology of Lakes and Ponds.

Although this book is an introductory text to aquatic ecology and limnology, we have provided additional opportunities for more in-depth study by referring to more specialized books in *Further reading*. Moreover, at the end of each chapter, we have suggested some practical aspects, including experiments, field studies, and questions for discussion. These features, together with the many illustrations and the *Glossary*, make the book suitable not only in traditional teaching, but also for independent study and, especially, in situations with only supervising teacher support, such as in problem-based learning (PBL).

We have written this book in close co-operation with each other and have had almost daily discussions over important as well as more cosmetic decisions during the writing process. We are therefore equally responsible for the final product and the order of authorship is alphabetic. However, writing a book is not something one does in isolation. We have had a lot of help from our friends and colleagues. Görel Marklund made all the beautiful line drawings, Marie Svensson and Steffi Doewes helped us with the layout, and Colin Little, Mikael Svensson, Staffan Ulfstrand, Lars Leonardsson, Gertrud Cronberg, Lars Tranvik, Wilhelm Granéli, Tom Frost, Larry Greenberg, Ralf Carlsson, and Stefan Weisner provided constructive comments on early manuscripts and the first edition. Several classes of

Limnology at the Department of Ecology, Lund University and Environmental engineering at Lund Institute of Technology, have given helpful and encouraging comments. We also thank our editor of this 2nd edition, Ian Sherman at Oxford University Press, who patiently and enthusiastically encouraged us throughout the writing process. The Swedish Research Council (VR) provided funds for the illustrations. Special thanks to Nelson Hairston Jr. and Sr. for allowing us to use '*Étude*'. Finally, we would like to thank our wives, Eva and Ann-Christin, and children, Victor, Oscar, Emilia, and Linn, Sigrid, Yrsa, for sharing the struggle for existence, as well as the joys and pleasures, of our everyday lives.

June 2004   C.B.
                                                                                                         L-A.H.

Contents

1 Introduction

If you go to a lake or a pond and make a thorough survey of the organisms, you will arrive at an impressive list of species, ranging in size from virus to fish. However, if you sample different kinds of lakes you will realize that the organisms in any one lake are only a small subset of all the freshwater organisms available. Further, you will see that some organisms will only appear in certain kinds of lakes or ponds, systems that provide very specific living conditions, whereas others inhabit a broad range of systems. The abiotic conditions in lakes may differ greatly; pH in freshwater systems, for example, may range from 2 to 14. Naturally, there are no organisms that have all the physiological or morphological adaptations needed to cope with such a breadth in an environmental gradient. Instead, organisms typically have adaptations that allow them to subsist within a narrower window of the abiotic variation. Thus, a lake provides an *abiotic frame*, an arena, made up of all the physical and chemical characteristics of that specific lake, such as morphology, sediment conditions, nutrient concentrations, light availability, pH, and temperature. This abiotic frame will differ spatially among, and even within lakes, as well as temporally within a specific lake or pond (Southwood 1988; Moss *et al.* 1994). Only organisms that have adaptations allowing them to survive under these conditions (i.e. their *niche* fits within the abiotic frame), will be able to colonize successfully and reproduce. To envisage this abiotic frame we could imagine a lake where only three abiotic variables are important, say light, temperature, and oxygen. If we plot the range of values these variables could have in this lake, we will arrive at a three-dimensional abiotic frame, a cube (Fig. 1.1) which defines the abiotic living conditions for colonizing organisms. In Chapter 2 (*The abiotic frame*), we present the most important abiotic processes of freshwater lakes and the adaptations that freshwater organisms have evolved to cope with these constraints. For practical reasons this frame is symbolized in two dimensions only (Fig. 1.2).

The third chapter of this book is a discussion about the *organisms* that can be found in lakes and ponds (symbolized by small circles, Fig. 1.2),

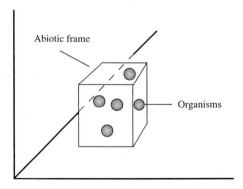

Environmental gradient

Fig. 1.1   A graphical representation of the abiotic frame (i.e. the environmental arena provided by a specific lake). Axes are different environmental gradients (e.g. light, temperature, oxygen). Spheres are species that have a niche that fits within the abiotic frame and thus may colonize this specific lake.

including taxonomy, morphology, and autecology of the more common organisms. Many of the organisms presented in this chapter are used in examples in other parts of the book and, hence, this will provide you with a profile of organisms occurring in lakes and ponds. Chapter 3 can also be used as a primary source when specific information about various freshwater organisms is needed.

Clearly, the abiotic frame excludes a lot of organisms that are simply not physiologically or morphologically adapted to survive in that specific system. On the other hand, organisms adapted to the abiotic characteristics of the lake or pond are free to remain and interact with other organisms. Why then do organisms not have the adaptations that would allow them to colonize a much larger range of lakes? Given the best of all worlds (Voltaire 1759), where all organisms actually had the evolutionary opportunity to acquire all adaptations, there still are allocation constraints that have to be considered. A game of cards can illustrate this: imagine that an adaptation can be represented by playing cards which the organisms can use in the everyday game of 'struggle for existence'. There is, however, not enough luck (energy) to have good cards for all needs—you cannot expect to get a hand with only aces and kings. Then imagine four areas within which it may be worthwhile to have good adaptations: being good at *gathering resources* may be one, let us call this adaptation spades; predator *defence adaptations* may be another, we can call this clubs; hearts is the obvious card symbol for being successful at *reproduction*, and adaptations to *withstand abiotic constraints* are given the symbol diamonds. We now have four suits, each symbolizing a set of adaptations to improve the possibilities of survival and reproduction. As concluded above, there is not enough luck (energy) to have a high score in all these domains, and we have to limit the total score of an organism's

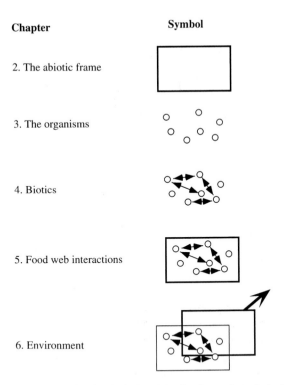

Chapter	Symbol
2. The abiotic frame	
3. The organisms	
4. Biotics	
5. Food web interactions	
6. Environment	

Fig. 1.2   Overview of the topics in this book depicted in the form of symbols. See text for explanation.

'hand' to 21 points. In an evolutionary perspective, this means that a species that uses most of the available points on being king of the lake at, for example, gathering resources (spades), will have mediocre cards in the other colours (Fig. 1.3). The development of the game will determine the success of this *specialist*; it is likely to become dominant during conditions when competition for resources is important and when protection, reproduction, and abiotic constraints are not major problems. However, many species are *generalists*, spending their 21 points more evenly among traits; being 'good enough' in most situations, but never superior (Fig. 1.3).

In a world of constant change, including seasonal variation and spatial differences within and among freshwaters, opportunities for specialist approaches will occur now and then, often leading to complete dominance within a limited area during a short period. Generalists, on the other hand, are seldom dominant, but are present almost everywhere and all the time.

Biotic interactions, including predation, competition, parasitism, and mutualism (symbolized by arrows in Fig. 1.2), determine the ultimate success of

Specialist Generalist

Fig. 1.3   Two hands of playing cards symbolizing a specialist and a generalist species, respectively. Each colour represents a specific adaptation. See text for further discussion.

any organism within the abiotic frame. Thus, the distribution and success of an organism is a function of abiotic constraints and biotic processes. The hierarchical importance of different factors can be illustrated with an example, the distribution of freshwater snails in waterbodies of increasing size (Fig. 1.4; Lodge *et al.* 1987; see also Wilbur 1984, for a similar model for the success of tadpoles in ponds). In their conceptual model, Lodge *et al.* (1987) argued that the availability of calcium is the most important factor determining the distribution of freshwater snails among lakes and ponds on a regional scale. Snails need calcium for building their shells and can only live in lakes with calcium concentrations greater than 5 mg l^{-1}; thus, this abiotic constraint sets the first restriction for the colonization of snails to a lake or pond (Fig. 1.4). In very small ponds other abiotic constraints are probably most important for snail populations. Temporary ponds do not hold water year-round and not all snails have adaptations that allow them to survive a dry period, and further, small, shallow ponds may freeze solid during harsh winters in temperate regions. In small but permanent ponds these abiotic disturbance events will not affect the population densities of snails; they will build-up to such high numbers that resources will become limiting and then competition, a biotic interaction, will be the most important structuring force (Fig. 1.4). Furthermore, in larger ponds and lakes, predators, such as crayfish and fish, will become abundant and these will reduce snail populations to levels where competition is no longer important; predation is then the most important structuring force (Fig. 1.4). Hence, the abiotic frame in this example is set by calcium availability and pond size. Once certain values of these variables have been reached, biotic interactions come into play. In Chapter 4 (*Biotics*), we elaborate on the importance of biotic interactions (competition, herbivory, predation, parasitism, and symbiosis), theoretical implications of these interactions, and how they affect the ecology of

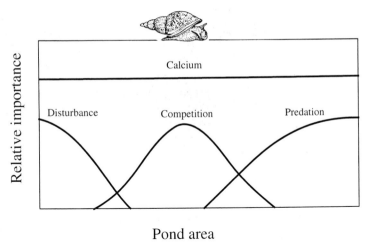

Fig. 1.4   Factors determining the distribution of freshwater snails in ponds of increasing size. Factors are ranked in relative importance. After Lodge *et al.* (1987).

organisms at the individual and population level. Competition between two populations that use the same resource or the effect of a predator on its prey may thus be important for the structure and dynamics of pond and lake systems. However, in natural systems it is not just two species that coexist and interact, but there are a multitude of predator, herbivore, and primary producer species and the interactions among all these species may be extremely complex. Indeed, it may at times seem hard to draw any conclusions on how these communities are structured and almost impossible to predict what will happen if the system is disturbed in any way. To make this somewhat overwhelming complexity a bit more easy to handle, species are often grouped into categories based on their trophic status; they are grouped into **trophic levels**, such as predators, herbivores, detritivores, and primary producers. We can then link these trophic levels into **food chains** or **food webs** and start to examine the direct and indirect interactions between organisms at different trophic levels. A number of more or less complex interactions between trophic levels have been identified and may indeed be of great importance in freshwater communities in both pelagic and benthic habitats. In Chapter 5 (*Food web interactions*), symbolized by a combination of the three symbols outlined above (Fig. 1.2), we consider the theoretical background and show some results of empirical tests of food chain theory from different habitats and with different organisms. The last section of Chapter 5 deals with how freshwater organisms can affect the abiotic frame they are living in. For example, recent research has shown that biotic processes may affect abiotic features of the lake, such as temperature and carbon availability.

The final chapter of this book deals with current environmental issues in freshwater systems (*Biodiversity and environmental threats*; Fig. 1.2).

Human-induced disturbances often alter the abiotic environment of a lake and we have chosen to symbolize this as a change in the position of the abiotic frame, with the result that some organisms fall outside the frame— they will disappear. We present the background to different environmental problems and the different measures that have been taken to accelerate the recovery of disturbed lakes and ponds. Issues discussed include eutrophication, acidification, ultraviolet radiation, and the introduction of exotic species, but also synergistic effects of these threats. We will also look closer at the effects small, as well as large-scale environmental changes may have on the **biodiversity** of organism communities. Although many environmental problems are severe and approaching disaster in several regions of the world, significant efforts are now being made to attain a more sensible use of freshwaters. It is hoped that this will result in more thorough planning and management of freshwater resources permitting a sustainable development of aquatic ecosystems.

2 The abiotic frame and adaptations to cope with abiotic constraints

Introduction

The abiotic characteristics, including physical and chemical factors, of a specific lake or pond constitute the frame within which the freshwater organisms live. Some organisms have adaptations that fit very well within this frame, others not so well, and still others not at all. This imaginary abiotic frame can be viewed as the sum of all physical and chemical characteristics of a specific lake or pond. This chapter covers the principal abiotic factors constituting the frame, including turbulence, temperature, pH, habitat permanence, as well as the availability of light, carbon, nutrients, and oxygen. Variations and temporal fluctuations in the abiotic frame among, as well as within, lakes and ponds call for different adaptations among organisms proposing to live in a specific waterbody. However, adaptations to cope with constraints posed by environmental features have no taxonomic borders, so that adaptations to survive at low oxygen concentration or pH are found in bacteria as well as in macrophytes and fish.

An organism's specific adaptations indicate under what environmental conditions it is likely to be successful and reproduce. Hence, knowledge of adaptations among organisms may be used to predict community composition in a certain environment. In the following pages we discuss different aspects of the abiotic frame from the organism's point of view (i.e. how different abiotic factors affect the performance of organisms). Since the title of this book includes both 'lakes' and 'ponds' it may be appropriate to note that the degree to which a waterbody is affected by abiotic features, such as wind-induced turbulence and temperature mixing, has actually been used to distinguish between lakes and ponds. A common definition of a **lake** is a waterbody in which wind-induced turbulence plays a major role in mixing of the water column, whereas in a **pond** more gentle temperature-induced mixing is the rule.

We will begin by looking at the problems of staying afloat and the importance of temperature in the functioning of freshwater ecosystems.

Staying afloat

Planktonic organisms generally have a higher density than water, which makes them sink through the water column and eventually end up at the lake bottom. Since they all want to avoid that, a multitude of solutions to this problem have evolved. Actually, almost anything you can imagine to reduce sinking can be found among planktonic algae and other organisms. In the following sections, we will provide an overview of the ingenious physiological and morphological adaptations these organisms have, including structures resembling parachutes, life-vests, flotation devices, and propellers!

Adaptations among algae—how to stay planktonic

Since most algae have a higher density than water, those with no adaptations to stay buoyant will only have a chance to dominate the phytoplankton community at turbulent conditions, such as during spring and autumn turnovers in temperate lakes. However, most algae have adaptations to reduce sinking rate by decreasing *density*, adjusting *size* and *shape*, or having *motility organelles*. Theoretically, the sinking velocity of a spherical particle follows Stokes' law:

$$v = \frac{2gr^2}{9n} (\rho - \rho_0) \text{ (Stoke's law)}$$

where g is the gravitational acceleration (m s^{-2}), n is the coefficient of viscosity of the fluid medium (kg m^{-1} s^{-1}), and ρ_0 is its density, whereas ρ is the density and r the radius of the particle (in this case the algal cell; units: kg m^{-3} and m, respectively). Accordingly, a small algal cell (low r) has a lower sinking rate than a large (high r), and, thus, in an evolutionary perspective, a small cell may not need additional adaptations to stay planktonic, whereas a large cell has an evolutionary pressure to solve the problem of how to remain in the water column. Therefore, the more sophisticated adaptations to stay planktonic are generally found in larger algae.

Density

Some algae can change their density (ρ), for example, by forming *vacuoles*, gas *vesicles*, and *ballast molecules*. At a high photosynthetic rate, heavy carbon products are formed which can be stored in ballast molecules, leading to increased density and, thus, increased sinking rate. When these heavy products are consumed during cell metabolism, density is reduced and the cell will again be positively buoyant. Other algae have gas vesicles which at low photosynthesis are filled with gas, making the cell more buoyant. When the cell reaches the surface, light intensity and thus photosynthesis increase, causing higher turgor (pressure) in the cell. This results in a collapse of the gas vesicles and reduced buoyancy. In this way the cell can optimize its position according to the availability of light (near the surface) or nutrients (which generally have higher concentrations near the sediment surface).

A layer of *mucus* covering the cell surface of many algae may also reduce sinking rates. Mucus is a complex polysaccharide, which itself has a high density but is able to hold large volumes of water such that the cell's average density becomes closer to water's. Thus, by excreting polysaccharide mucus, the cell increases its cell size, but reduces the average density and thereby the sinking rate.

Size and shape

Another option, not covered by Stokes' law, which only applies to spherical particles, is to have a shape that reduces sinking. Any shape differing from a sphere will reduce sinking. The most common morphological adaptation to reduce sinking in, for example, diatoms and cyanobacteria, is to form *colonies*. Alternatively, different types of *spines* increase the surface area in the same way as a parachute. However, it is still under debate whether some types of spines really reduce sinking. For example, the spines of *Staurastrum* (Fig. 3.7) are sometimes covered with a mucous layer resulting in a spherical shape of the algae to which the spines only add weight. But some algae, such as *Scenedesmus* (Fig. 3.9), have delicate spines that really do reduce sinking rate. Spines also cause the cell to tumble through the water, which may, besides reducing the sinking rate, be an adaptation to facilitate nutrient uptake and evacuating excretion products (Hutchinson 1967).

Flagella

A rather sophisticated adaptation to stay planktonic is the use of *flagella*, which not only permit the cell to adjust its position vertically, but also allow movement in any direction. A flagellum can work either as a propeller or as a whip moving back and forth, thereby making the cell move forward. Some flagella-carrying algae, such as *Gymnodinium* (Fig. 3.6), may reach speeds of up to 1.8 m h^{-1} (Goldstein 1992).

Turbulence

Together with temperature, exposure to wind creates turbulent currents making the aquatic environment far more homogeneous, than terrestrial habitats. Wind-induced turbulence is, of course, more influential in large, wind-exposed lakes than in small, sheltered forest ponds. Wind may also create various phenomena in lakes, such as the spiral-formed, subsurface

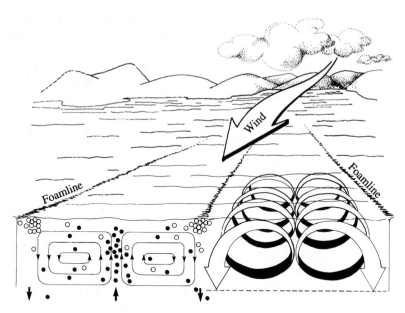

Fig. 2.1   A cross-section through a lake showing foamlines caused by wind-induced rotations of the water column—Langmuir rotations. Owing to the rotations, positively buoyant particles (open symbols) will be accumulated in surface foamlines and negatively buoyant particles (closed symbols) in the upwelling current between two rotations. Partly based on George (1981) and Reynolds (1984). Reproduced with permission from the Freshwater Biological Association.

turbulence called **Langmuir rotations**. The visible result of these rotations, which can be seen in any lake, are surface foamlines that run parallel to the wind direction (Fig. 2.1). Positively buoyant particles and small organisms tend to concentrate in the foamline at the surface, negatively buoyant objects in the upwelling current between two rotations, whereas objects with neutral buoyancy tend to be evenly distributed in the water column. Phytoplankton and zooplankton are often concentrated to patches between rotations and organisms as large as phantom midge larvae (*Chaoborus flavicans* Fig. 3.20) may be affected by the wind-induced water circulations. A study of the diurnal distribution patterns of *Chaoborus* using echosounding technique revealed that the upwelling water in the Langmuir rotations 'snatched' chaoborids from the deep, oxygen-poor water where they spend daytime hours to avoid predation by fish (Malinen *et al.* 2001). The Langmuir rotations concentrated the chaoborids into clouds in the upwelling area and these clouds attracted smelt, the dominant planktivorous fish in the lake. Thus, wind-driven rotations resulted in changes in the distribution patterns of this invertebrate, driving it out of its refuge and exposing it to predation. Similarly, on a whole lake scale, wind may concentrate positively buoyant objects at the surface of the windward side of the lake, whereas negatively buoyant objects will be concentrated at the leeward side.

Another visible phenomenon induced by strong winds is **seiches**, where lake water is 'pushed' towards the windward side of the basin. This may increase the water level significantly at the windward beach of large lakes and decrease it at the leeward beach. When the wind stress is removed, the water level 'tilts back' and a standing wave is created that moves back and forth. Another type of seiche is the internal, or thermocline, seiche, which is also induced by wind creating a surface current. When this current reaches land it bends down and thereby presses the surface water package, including the **thermocline** downwards (see section 'Temperature basics'). Since the thermocline often functions as a 'lid' between cold, oxygen-depleted, and nutrient-rich bottom water and warm, oxygen-rich, but nutrient-depleted surface water, these seiches may be very important for the exchange of heat, oxygen, and nutrients between the two, relatively isolated, water layers. The amplitude of internal seiches may be several metres and they may rock back and forth for several days, injecting oxygen to the bottom water and causing upwelling of nutrients to surface water to the benefit of phytoplankton. Although these large-scale processes affect the everyday life of organisms in the lake, there is little a specific organism can do except to surrender to the enormous forces induced by wind. Turbulence may, however, have positive effects for the everyday life of many organisms.

Positive effects of turbulence

Many organisms in streams and rivers are supplied with nutrients or food particles by the running water; they can just sit and wait for the food to arrive! This is generally not the case in lakes and ponds and if, for example, an algal cell is not in motion, a microlayer that is depleted of nutrients, but have high concentrations of excretion products, is formed around the cell. By being in motion, as a result of turbulence or an irregular shape causing the cell to tumble around, it avoids such microlayers which may otherwise severely deteriorate cell metabolism. Turbulence is even more important for periphytic algae, living their lives attached to a substrate. If they happen to be attached to an inert substrate that does not leak any nutrients, such as a stone or a car wreck, they have to rely entirely on nutrients and carbon dioxide provided by a continuous flow of water. Thus, turbulence may be of utmost importance for nutrient supply and removal of excretion products.

Lake formation

Different lake morphologies affect physical properties such as turbulence, circulation, temperature gradients, and stratification, but also give rise to different levels of productivity, suggesting that the processes that have formed a lake may be of considerable importance also for its biology.

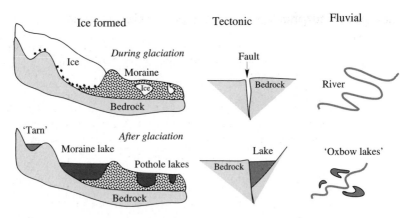

Fig. 2.2   Examples of lake types formed by the retreat of glacial ice, including 'tarns', moraine lakes, and pothole lakes, and lakes formed by tectonic movement of the bedrock, as well as oxbow lakes formed from changes in the flow of rivers. The upper panel shows the situation before the lakes were created, whereas the lower shows the present situation.

Therefore, we will glance at some of the ways freshwater systems were, and in some cases still are, formed.

Some of the oldest and largest lakes were formed by **tectonic** activities; that is, various types of movements of the earth's crust. Examples of such lakes are the African Rift Valley lakes and the deepest lake on earth—Lake Baikal in Russia (Fig. 2.2). Especially in temperate and polar regions the retrograde movement of the glacial ice shelf resulted in irregularities in the landscape, which were then filled with water. An example of such **ice-formed lakes** is when glaciers were retreating and some large ice blocks were left behind. When these blocks eventually melted basins were formed which were soon filled with water. Such lakes are called kettle- or pothole lakes (Fig. 2.2). A glacier is extremely heavy and also changes shape and size over time as a result of freezing and thawing cycles, scouring the underlying bedrock and thereby incorporating and transporting huge amounts of large and small stones; from boulders to particles smaller than sand grains. The weight and scouring created depressions and when the ice melted off, these depressions were filled with water, creating lakes called 'tarns' (Fig. 2.2). Moreover, at the end of the glacier, a lot of rocks and sand were deposited, forming a wall—a terminal moraine—which prevented water from flowing downstreams and, voilá, a moraine lake was formed! (Fig. 2.2).

Another common way of lake formation, which does not have anything to do with tectonics or glacial activities, is when a river bend pinches off, leaving an often banana formed part of the former river behind. Such **fluvial** lakes and ponds are often called oxbow lakes (Fig. 2.2). There are, of course, other ways lakes and ponds can be formed, not to mention **man-made** reservoirs, but the tectonic-, ice-, and fluvially formed systems are, indeed, the most common ways of pond and lake formation.

Lake basin morphology—and our perception thereof

When a lake is finally created, either through tectonic, glacial, fluvial, or other forces, it has different morphometric features, such as surface area (A), maximum depth (z_m), mean depth (z), and volume (V), which is generally illustrated as in Fig. 2.3. The mean depth (z), can be roughly calculated as the volume divided by the surface area of the lake. On bathymetric maps, the **depth isoclines**, which are lines connecting sites with the same depth, are illustrated in the same way as mountains and hills are on terrestrial maps. Hence, where isoclines are close to each other the depth is rapidly increasing, as shown to the left in Fig. 2.3 (a), whereas a smooth depth increase is illustrated with long distances between isoclines as to the right in Fig. 2.3 (a). The shape of a lake or pond may vary from completely circular to an irregular shape with plenty of bays and points. A way of expressing the degree of irregularity is the shoreline development factor (D):

$$D = \frac{L}{2\sqrt{\pi \cdot A}}$$

where L is the shore length, A is the surface area of the lake. A perfectly circular lake or pond would then get a shoreline development factor of 1, whereas one with a complicated and long shoreline will have a $D > 1$.

When asked to draw a picture of a lake, most of us end up with a drawing of something like Fig. 2.3(a), that is, proportions resembling a bowl or a deep plate. However, very few lakes have such proportions and we tend to exaggerate depth. This can be illustrated with the proportions of the world's deepest lake, Lake Baikal in Russia (maximum depth, 1740 m; length, 750 km). A drawing to scale of Lake Baikal looks like a thin line even though the lake is almost 2 km deep (!) (Fig. 2.3(b)). The same is, as we see in the same figure (Fig. 2.3(c)), true for Lake Superior in the USA (maximum depth, 307 m; length, 600 km). We come to a similar conclusion if we just pick out an arbitrary example of an ordinary lake, let's say one that is 400 m long and 10 m deep, which results in a line that is 2 cm long and 0.5 mm thick!

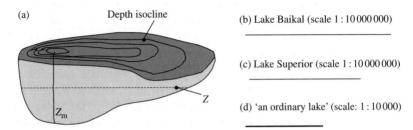

(a)　　　Depth isocline

(b) Lake Baikal (scale 1 : 10 000 000)

(c) Lake Superior (scale 1 : 10 000 000)

z

(d) 'an ordinary lake' (scale: 1 : 10 000)

z_m

Fig. 2.3　(a) Morphological features of a lake including area (A), volume (V), maximum and mean depths (z_m and z, respectively); (b-d) The lines represent the real proportions between length and depth of lakes, exemplified with some of the world's deepest lakes (Lake Baikal, Russia and Lake Superior, USA), as well as an 'ordinary' lake. Note different scales.

Hence, our perception of lakes leads to that we intuitively tend to believe that processes in the water column are the most important. However, this simple exercise may make it obvious that we should, when taking into account the real surface to volume relations, put more effort into understanding how processes in the sediment may affect the relatively small column of water above. We will come back to bottom (benthic) related organisms and processes later and also see how benthic and pelagic (open water) processes interact. But first, we will consider another feature that is strongly affected by lake morphology—the temperature.

Temperature

Temperature has traditionally been recognized as a key environmental factor in freshwater ecosystems, affecting, for example, distribution patterns, behaviour, and metabolic rates of freshwater organisms. The main difference between water and other media is its *high heat storage capacity*. Diurnal and seasonal temperature changes are therefore much less drastic for aquatic than for terrestrial organisms. Despite water temperature changes being more even than in terrestrial systems, vertical as well as seasonal changes in temperature are major determinants for the distribution, behaviour, and reproduction of organisms. The main reason for this is that most freshwater organisms are **poikilothermic**, which means that their internal temperature varies with environmental temperature and, thus, environmental temperature is a very important factor in their life. Before we investigate further temperature effects on the everyday life of organisms, some basic features of temperature need to be understood.

Temperature basics

The water molecule

Water's high capacity of storing heat; that is, its *high specific heat* (the energy it takes to heat 1 g of water by 1°C), is a consequence of its molecular structure. The water molecule is electrically asymmetric and this results in a strong electrostatic attraction, or hydrogen bonding, between the negatively charged oxygen atom of one water molecule and the positively charged hydrogen atom of an adjacent water molecule. The hydrogen bonds connect the water molecules in ice and water, forming a continuous network of water molecules. The strong hydrogen bond is the reason that water is a liquid at room temperature, otherwise it would be a gas. In ice, the water molecules form a rigid hexagonal crystal structure, a more spacious molecule structure than in water (Fig. 2.4) and, hence, ice has a density that is lower (0.917 g cm^{-3} at 0°C)than liquid water—ice floats on water! As it melts, some of the crystalline organization is retained, but molecules are allowed to move more freely and

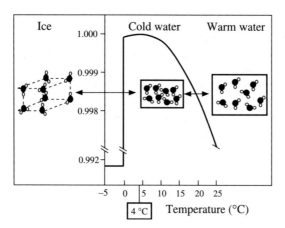

Fig. 2.4   The density of water along a temperature gradient. At temperatures lower than 0°C, the density is lower than 1.000 and the water molecules are arranged in a crystalline structure—ice. At about 4°C the density of water is at its maximum (i.e. the molecules are as closely packed as possible). As the temperature rises, the spaces between molecules increase, leading to a decrease in density.

collapse into spaces, increasing the number of molecules packed within a given volume (Fig. 2.4). This results in an increase in the density of water with increasing temperature. However, at the same time there is an increase in the kinetic energy of the water molecule and this affects water density in the opposite direction; increasing kinetic energy tends to increase molecular distances and decrease density. Thus, there is a peak in water density at 4°C (Fig. 2.4). As indicated above, these changes in density with temperature have profound effects on the spatial and temporal distribution of heat, and therefore on the performance of freshwater organisms.

Distribution of heat—temporal and spatial patterns

Although the variability of temperature in aquatic habitats is less than in terrestrial ones, freshwater organisms still experience temporal and spatial variations in temperature. Solar radiation is the major source of heat in lakes and ponds and, consequently, there is variation in water temperature both on a seasonal and a daily basis. Most of the incoming light energy is converted directly into heat. The absorption of light by water, dissolved substances energy and suspended particles is exponential with depth, so that most is absorbed within the first few metres. Wind-generated currents will distribute heat within the lake but only down to a limited depth. Thus, two layers of water will be formed, a warm less dense layer at the surface and an underlying layer of dense, cool water. This is called **thermal stratification** and is crucial for the physical, chemical, and biological processes in lakes. Because of changes in solar radiation and wind turbulence over the years, the stratification is not permanent, but varies between seasons. For a given lake, though, there is a highly predictable pattern in temperature distribution between years. Below, we will go through an annual cycle of a typical temperate lake of moderate depth, starting with ice-break, which for most organisms signals the start of a new growing season.

Stratification and circulation

As atmospheric temperature increases in spring, the ice cover will gradually decrease in thickness and eventually break up. At this time, the water temperature will be around 4°C (i.e. water will be at its maximum density). A few degrees difference in temperature with depth will not have any substantial effect on water density as the change in density is relatively limited within this temperature range (Fig. 2.4). Given the similarity in density throughout the water mass there will be little resistance to mixing and only a small amount of wind energy will be needed to mix the whole water column. Thus, as the lake loses its protective ice cover in early spring, the entire water mass will circulate, a process driven by wind energy and convection (temperature-induced) currents (Fig. 2.5). The lake water will circulate for some time, but as spring proceeds the intensity of solar radiation will increase and with it water temperatures. Because the main part of the heat content of light is absorbed in the first few metres of the water column, the surface water will be heated more rapidly. During a period of warm, calm weather an upper stratum of warmer water will develop. Given the exponential decrease in density with temperature there will be a relatively significant difference in density with only a few degrees difference in temperature between the warmer surface layer and the colder bottom layer. This difference in density between the two layers will be sufficient to resist mixing as windier weather resumes. Thus, the lake has been stratified into two layers, the upper, warmer **epilimnion** and the underlying water mass of cool, dense water, the **hypolimnion** (Fig. 2.5). The stratum between the two layers, characterized by a steep thermal gradient, is called the **thermocline** or the **metalimnion**.

In autumn, the energy input from solar radiation will decrease owing to the decreasing solar angle and at the same time the heat loss by evaporation will be continuously high. Thus, the epilimnion will have a net loss of heat, resulting in a decreasing temperature difference, and hence density difference, between epilimnion and hypolimnion. Autumn weather is characterized by increasing windiness and eventually, when the density differences between the strata are small enough, the wind energy will cause the whole water column to re-circulate; the autumn turnover (Fig. 2.5).

In winter, when the temperature of the lake has cooled down to about 4°C and water density is at its maximum, an inverse stratification will establish where the upper stratum has the lowest temperature. This is because water density decreases at temperatures below 4°C (Fig. 2.4) resulting in a layer of cool, less-dense water overlying dense 4°C water (Fig. 2.5). However, the density difference between these layers is only minute and thus will be easily disrupted by wind-induced water movements. This pattern will continue

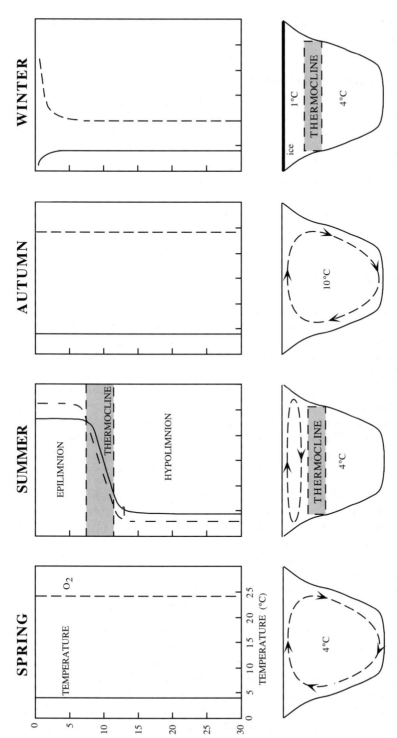

Fig. 2.5 Seasonal circulation of water, as well as depth gradients of temperature (solid line) and oxygen (dashed line) in a temperate lake. During stratification, the water column separates into three different layers with little exchange of water and nutrients. The metalimnion (thermocline) is the layer where temperature changes drastically. The layers above and below the metalimnion are the epilimnion and the hypolimnion, respectively.

until surface ice can form, typically during a cold night with no wind, and seal the lake from further wind energy impacts.

The depth of the epilimnion and the temperature of the hypolimnion will depend on a number of factors, such as latitude, weather, water clarity, and lake morphometry. Because most of the transport of heat to deeper levels is due to wind action, characteristics of the lake that may modify the wind effect will be important for the formation and depth of the thermocline. *Lake surface area, mean depth, volume, exposure,* and *fetch* (the longest distance of the lake that the wind can act on uninterrupted by land) will all affect how efficiently wind energy can transport heat to deeper strata. The biotic structure of the lake may also influence the depth of the epilimnion (Mazumder *et al.* 1990; Mazumder and Taylor 1994) through changes in water clarity. The depth of the epilimnion was greater at low than at high fish abundances. The probable explanation is that at low fish abundances, large zooplankton thrive and reduce the biomass of phytoplankton and thus increase water clarity (see Chapter 5, *Food web interactions*).

The temporal pattern in temperature, including formation of the thermocline, is the most important physical event for the structure and function of the lake, and will dramatically affect the conditions for the lake biota. After stratification, the lake will be divided into two separate compartments, a warm, circulating upper layer with high light intensity and where the major part of primary production takes place, and a cold, relatively undisturbed layer where the decomposition of organisms sedimenting from the epilimnion takes place. Due to the drift of dead organisms into the hypolimnion and the limited exchange between the two strata, nutrient availability will soon be a limiting factor for primary production in the epilimnion.

Superimposed on seasonal changes there are diurnal cycles in temperature. Diurnal changes in water temperature are most prominent in shallow habitats, such as in weed beds of the littoral zone, and in general, the smaller the volume of the waterbody the greater the amplitude of temperature an organism will experience on a diurnal basis.

Types of mixing   The thermal pattern we have described above is typical for temperate lakes of moderate depth that have ice cover in winter. These lakes are called **dimictic** because they mix twice a year. Obviously, there exist differences in stratification patterns between lakes depending on, say, local or regional differences in climate, lake morphometry, and exposure to wind. **Monomictic** lakes, for example, are never ice-covered and circulate all through the winter. In shallow, wind-exposed lakes in warmer regions the stratification may be quite unstable, only lasting a few weeks at a time. Such lakes circulate many times a year and are called **polymictic**. Lakes may also differ in the extent of mixing during the mixing cycle. In **holomictic** lakes the whole water column is mixed during circulation, whereas in

meromictic lakes, typically lakes with a large depth, the bottom layers will not be involved in the mixing.

Adaptations to fluctuations in temperature

Physiological processes, such as metabolism, respiration, photosynthesis, but also activity patterns and behaviour are all temperature-dependent. As we have seen above, freshwater organisms live in an environment that is relatively stable considering temperature fluctuations, but they still experience a wide range in temperature, varying both seasonally and diurnally. The rate of physiological reactions taking place within an organism is determined by enzyme systems that are temperature-dependent and have an optimal temperature, on either side of which they function less efficiently. Too high temperatures may result in enzyme inactivation or even denaturation. The temperature optimum is not the same for all organisms, but, rather, different organisms have different temperature optima. Fish, for example, are often separated into 'cold' and 'warm' water species based on temperature preferences estimated in the laboratory and from distributions in the field (e.g. Magnuson *et al.* 1979; Fig. 2.6). The width of the temperature range within which an organism can survive and reproduce also differs between species; some species can tolerate a wide range of temperatures

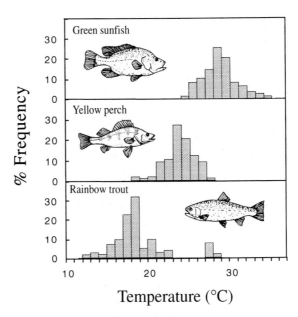

Fig. 2.6 Temperature preference for warm, cool, and cold water fish as determined in laboratory preference experiments, exemplified with green sunfish, yellow perch, and rainbow trout, respectively. The histograms show the percentage of time spent at each temperature. From Magnuson *et al.* (1979).

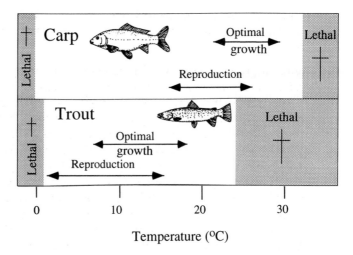

Fig. 2.7   Thermal requirements of carp and brown trout, showing the temperature ranges within which each species can survive, reproduce, and where they have their optimal growth. Modified from Elliott (1981).

and are called **eurytherms**, whereas others only thrive in a narrow temperature window—**stenotherms**. A species' temperature range can be defined in different ways: as the range between cold and warm lethal temperatures, the range an organism prefers in laboratory experiments, or the range of temperatures where it can be found in natural habitats. In the upper and lower limits of a species' temperature distribution, the metabolic costs of surviving are so high that no energy can be allocated to other functions, such as growth and reproduction. Typically, mere survival occurs over a wide temperature range, whereas growth only takes place in a more limited temperature range. Optimal growth and reproduction occur in an even more restricted range (Fig. 2.7).

With respect to many other organisms, such as algae, a rule of thumb is that maximum growth rate doubles with every 10°C increase in temperature. Different algal species have different optimal growth temperatures, which for most algae are between 15° and 25°C, with very low growth rates below 5°C and above 30°C, the latter mainly due to destruction of the enzymes of cell metabolism.

Acclimatization

Temperature tolerances are not necessarily fixed, but can change if the organism is exposed to different temperatures; organisms can **acclimatize** to a changing temperature regime. An example of this is three algal strains of the same green algal species (*Chlorella sp.*) that showed maximum growth rates at 20°, 26°, and 36°C, respectively, depending on the temperature in which they had previously been cultured (Yentsch 1974). Hence, acclimatization caused a deviation in optimal temperature of 16 °C! Similar acclimatization to regional temperature regimes is common among many other organisms.

Some algal species are adapted to extreme temperature regimes; found in, for example, snow or hot springs. Such habitats offer the advantage of *reduced competition* and, usually, a complete *release from grazing*. The most common species of snow algae is *Chlamydomonas nivalis* (*Protococcus nivalis*), having resting cells containing the pigment astaxanthin, which can make entire snow fields look bright red. Hot springs offer another extreme environment devoid of grazers and with a high supply of nutrients. The most tolerant organisms are certain species of bacteria and cyanobacteria, which are able to grow at 70°C, and may survive at temperatures close to the boiling point. Diatoms and chlorophytes ('green algae') are less tolerant, but are able to grow at temperatures close to 50°C. On the other hand, very few animals are able to survive at such high temperatures, but some species of copepods have been found in hot springs at temperatures of 38–45°C.

Resting stages

At suboptimal conditions (e.g. low temperatures), several organisms form resting stages, which, among algae, generally are thick-walled cysts that are morphologically different from vegetative cells. Cysts are able to withstand various adverse environmental conditions including suboptimal temperatures, anoxic conditions, exposure to fungal and bacterial growth, and almost dry conditions since they may remain viable with the help of pore water trapped between sediment particles.

Not only algae, but also some zooplankton species form 'egg banks' at the sediment surface, and just as for algae, the eggs will hatch in favourable conditions. Exactly which the specific factors are that induce hatching is largely unknown, but internal factors such as genetically programmed timing, and external factors such as physical and chemical conditions are probably involved. In a study where the egg viability of the copepod *Diaptomus sanguineus* was determined (Hairston *et al.* 1995), eggs as old as 332 years were hatched in the laboratory and viable copepods emerged. Hence, these resting eggs were deposited when European settlers had just started their colonization of North America!

The zooplankton, *Daphnia* also forms resting eggs that are released into a protective chamber called the **ephippium**, which can be seen as a black 'back-pack' on females (Fig. 3.13). The formation of ephippia may be induced by low temperature, shorter photoperiod, low food supply, or even by the presence of predators (Slusarczyk 1995, Gyllström and Hansson 2004). Eggs in ephippia generally hatch within a few months, but may rest longer if the conditions are not favourable enough.

An egg or cyst bank functions as a time-dispersal system that allows rapid recolonization following a catastrophic event. Eggs and cysts may be buried deep into the sediment, probably due to the activity of benthic animals, and may stay there until a fish or a limnologist's anchor disturbs the sediment and the egg or cyst is exposed to favourable conditions. A striking example of how sediment disturbance can affect the phytoplankton community is demonstrated by the restoration of Lake Trummen (Sweden). Large amounts of surface sediments were removed and exposed resting cysts of chrysophycean algae to more favourable conditions. This resulted in a dramatic increase of these algae in the lake water (Cronberg 1982).

Temperature-induced hatching

One of the most important effects of temperature on poikilothermic organisms is its effect on the rates of development and growth. For example, eggs of the fish species roach (*Rutilus rutilus*, Fig. 3.21) do not hatch at the same time every year. Rather, hatching date is dependent on the accumulated temperature during the period preceding hatching (i.e. the sum of **degree-days**). Thus, hatching time varies between years due to between-year variability in weather, but organisms will always hatch after a fixed degree-day threshold. The availability of food resources and/or the activity of predators will also be dependent on temperature, and the timing of reproduction must ensure that the larvae are hatching at a time when resources are abundant so that they can grow rapidly to a size where they are less vulnerable to predation.

Behavioural thermoregulation

As we have seen above, there is a spatial variation in temperature within lakes and ponds and this is especially pronounced during the warmer season when most of the primary and secondary production takes place in freshwater systems. Many freshwater organisms thus have an opportunity to change their thermal surroundings by moving to a different microhabitat, i.e. behavioural thermoregulation. One of the most spectacular patterns of movement in freshwater organisms is the diel vertical migration displayed by many groups of organisms, such as phytoplankton, zooplankton, and fish. Migrating organisms typically spend their days in the cold hypolimnion and move up into the warmer epilimnion only during the night. One hypothesis explains diel vertical migration as a behavioural thermoregulation, where migrating organisms could forage at high rates in the warm epilimnion and conserve energy (reduce metabolism) by migrating to the cold hypolimnion. However, recent research on zooplankton diel vertical migration has shown that the main benefit of downward migration at dawn is to reduce predation by visually orienting predators (see also Chapter 4, *Biotics*). Nevertheless, examples of behavioural thermoregulation may be found among fish, such as the sculpin, *Cottus extensus*, which has a thermoregulatory strategy to increase digestion and growth in juveniles (Neverman and Wurtsbaugh 1994). Juvenile sculpins spend the day feeding on benthic organisms in the cool (5°C) hypolimnion of the lake and migrate 30–40 m into the warmer (13–16°C) metalimnion or epilimnion at night (Fig. 2.8). It appears that food is not limiting for juvenile sculpin growth, but rather that the digestion rate is low in the cold hypolimnion. In the warm epilimnion digestion is rapid and juvenile growth rate is high. Thus, juvenile sculpins fill their stomachs during the day feeding in the cold bottom layers and migrate up to warmer water

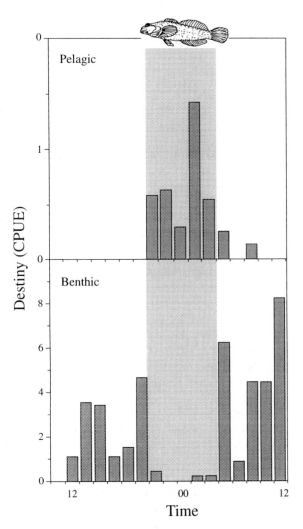

Fig. 2.8   Diel catch rates of juvenile sculpins in benthic and pelagic trawls measured as catch per unit effort (CPUE). Note that juvenile sculpins spend the day feeding close to the bottom and then migrate up to warmer water for digestion in order to increase growth rates. The shaded area denotes dark hours. Modified from Neverman and Wurtsbaugh (1994).

during the night to digest the meal. This increases juvenile growth rates and thereby decreases the time they are vulnerable to predation from gape-limited predators.

Interactions between temperature and other factors

As we have seen, temperature is a very important environmental factor that has profound effects on physicochemical as well as biological processes in lakes and ponds. However, when considering the effect of temperature on biological processes it is sometimes hard to separate the effect of temperature from that of other factors as they often covary in time and space. Light, which follows a similar seasonal cycle as temperature, and oxygen, which dissolves better in water at lower temperatures, can serve as two examples of covariation with other factors. Further, biological processes may be governed by two or more different factors that act in concert or against each other. For example, the amount of light that is needed to saturate photosynthesis of the submerged macrophyte *Potamogeton pectinatus* decreases with increasing water temperatures (Madsen and Adams 1989). Such interactions between abiotic factors lead us directly into another important axis of the abiotic frame—light.

Light

Solar energy provides the major energy input to ponds and lakes. The radiant energy of light can be transformed into potential energy by biochemical processes such as photosynthesis. Light is also absorbed and transformed to heat by particles, dissolved substances and by water itself. Absorption of light by water is high in the infrared region of the spectrum (>750 nm), decreases in the visible part (750–350 nm), and again increases in the ultraviolet (UV; <350 nm) region of the spectrum due to interference with water molecules (Fig. 2.9). In pure water, all the red light is absorbed in the first few metres, whereas almost 70% of the blue light remains at 70 m depth (Fig. 2.9). Generally, more than 50% of the total light energy is absorbed in the first metre of the water column. The depth to which light can penetrate is mainly determined by the amount of dissolved and suspended substances in the water. Light extinction in a uniform water column is exponential, which means that light is reduced by a fixed proportion at each depth increment. If the percentages of surface light remaining at successive depths are plotted against depth, an exponential curve is obtained (Fig. 2.10), but if, instead, light intensity is expressed on a logarithmic scale, the decrease in light intensity with depth is described as a linear function. The main reason for transforming light intensity into a logarithmic scale is that it simplifies the determination of the depth where

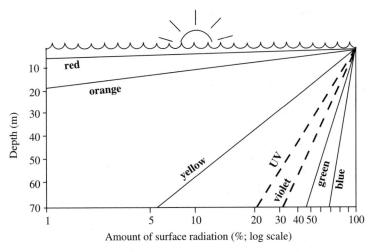

Fig. 2.9 Light transmission in distilled water showing that red light (680 nm) is already attenuated at a few metres depth, followed by orange (620 nm) reaching less than 20 nm depth. About 5% of the yellow (580 nm), 46% of the green (520 nm), and almost 70% of the blue (460 nm) light still remains at 70 m depth. Note that violet and UV radiation, which have the shortest wavelengths (400 nm and <350 nm, respectively) do not reach as deep as green and blue radiation.

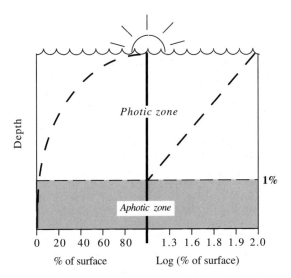

Fig. 2.10 Light extinction through the water column. The left curve shows light intensity as per cent of surface values, whereas the right curve (line) shows the same value but on a logarithmic scale. The slope of the line is the extinction coefficient. The volume where light intensity is above 1% of the surface value is the photic zone (i.e. where photosynthesis is higher than respiration). Below about 1%, in the aphotic zone, respiration consumes more oxygen than is produced by photosynthesis. The border between the two zones, where photosynthesis equals respiration, is called the compensation depth.

1% of the surface light remains. The 1% level is simply the intercept of the linear function, which is biologically important since it is used as an estimate of the maximum depth where net photosynthesis can occur. The part of the water column where there is a net oxygen production by primary producers (i.e. where light intensity is above about 1%) is called the **photic zone**, and the region below is called the **aphotic zone** of the lake. The border between these zones is where photosynthesis equals respiration—the **compensation depth** (Fig. 2.10).

Photosynthesis

The energy in light is used by plants in photosynthesis to build up large, energy-rich molecules from carbon dioxide and water. In higher plants (**eukaryotes**), the mechanism specialized for obtaining energy from sunlight is concentrated in the chloroplasts, which contain several pigments. In the primitive cyanobacteria, which lack chloroplasts, the pigments are distributed throughout the cytoplasm. Pigment concentrations and composition vary among algal groups, but all freshwater algae (including cyanobacteria) contain chlorophyll *a* and, beta-carotene. Chlorophyll *a* has its absorption maxima at 430 and 665 nm, and is the main pigment that uses solar energy to transfer carbon dioxide and water into sugar and oxygen:

$$6CO_2 + 6H_2O \rightarrow \boxed{\text{Light}} \rightarrow C_6H_{12}O_6 + 6O_2$$

Several accessory pigments are also active in the photosynthetic reaction, including beta-carotene, xanthophylls, and chlorophyll *b*. These pigments have light absorption maxima at other wavelengths than chlorophyll *a*, allowing a broader spectrum of wavelengths to be used by the plant. At changing light regimes, algae may optimize photosynthesis by adjusting the amount of chlorophyll in relation to the light intensity (i.e. at low light availability the chlorophyll content of the cell increases). Despite the fact that the chlorophyll *a* content of an algal cell may vary with light intensity, it generally constitutes 2–5% of the cell's dry weight. Owing to this relatively constant portion of dry weight among algae, and to simple and reliable methods for analysis, chlorophyll *a* is widely used as an estimate of algal biomass.

Measuring photosynthesis along a depth gradient during a day with full sunlight shows reduced photosynthesis near the surface due to **photoinhibition**, which is a well-known but not completely understood phenomenon.

Adaptations to optimize light acquisition

The amount of light affects the rate of photosynthesis, which in many planktonic algae, such as the cyanobacteria *Planktothrix* and *Anabaena*, leads to increased cell turgor and a collapse of labile gas vesicles, causing the alga to sink. When it sinks, the photosynthetic rate decreases, causing the vesicles to increase in size and the alga to become buoyant again. In this way, the amount of light determines the position of the algae in the water column. Except for shallow regions, the amount of light reaching the sediment surface is generally low. However, at sufficient amounts of light the sediment surface is usually covered with a thin biolayer of algae (**periphytic** algae). The obvious advantage of growing at the sediment surface is the endless supply of newly mineralized nutrients. However, the sediment surface is in constant motion due to water movements and animal activity, exposing the sediment-dwelling algae to a constant risk of being buried. As an adaptation to this environment, some algal groups, especially diatoms, are able to perform gliding movements. In combination with **phototactic** behaviour (movement towards the light), gliding movements make it possible to avoid being buried alive in the complete darkness of the sediment.

Macrophytes are, besides alga, the only aquatic organisms that need light as their energy source. Since light intensity rapidly attenuates through the water column, light availability will restrict submerged macrophyte growth to shallow depths. Angiosperms have been found only at depths of less than 12 m, whereas bryophytes ('mosses') and charophytes may grow at depths beyond 100 m in clear water lakes (Frantz and Cordone 1967). As light attenuation increases due to increasing particle concentration or changes in water colour the maximum depth where macrophytes can grow (Z_c) will decrease. This results in a positive relationship between Secchi depth and Z_c (Fig. 2.11; Chambers and Kalff 1985). The shape of the relationship and the maximum Z_c differs between macrophyte groups reflecting their light requirements. Angiosperms need most light to grow followed by charophytes

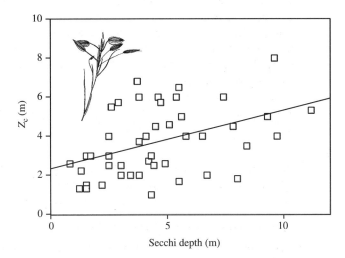

Fig. 2.11   The relationship between Secchi depth and maximum depth of growth (Z_c) for freshwater angiosperms. Data from Chambers and Kalff (1985).

and bryophytes. The average light requirements at Z_c, calculated as the percentage light remaining of the light at the surface were 2%, 5%, and 15% for bryophytes, charophytes, and angiosperms, respectively (Middelboe and Markager 1997). Although light levels set the lower depth limits where plants can grow in lakes, plants may escape light limitation by growing tall, thus reaching the better light conditions near the surface while still being rooted at depths where there is not enough light for photosynthesis. Perennial plants, that die down during winter, need to store energy, for example, in below-ground storage structures (tubers), to use for the first growth in early spring. Emergent macrophytes, however, have 'escaped' from the problem of light limitation by photosynthesizing above the water surface.

Abiotic factors affected by the catchment area

Temperature-, light-, and wind-induced turbulence are large-scale abiotic factors characteristic for whole regions. However, many other abiotic factors, including acidity, humic content, carbon and nutrient concentrations, and, indirectly, oxygen concentration, are merely results of characteristics on a much smaller scale—the **catchment area** of the lake or pond.

The catchment area

The catchment (or drainage) area of a lake is defined as the region around the lake or pond that drains the rain to the lake. The boundary of a catchment area is the ridge beyond which water flows in the opposite direction, away from the lake. The size, bedrock, soil, and vegetation of the catchment area affects the nutrient input, pH, and the water colour of the lake. In fact, the catchment area has a crucial role in shaping the chemical composition of the lake water. For example, when rainwater finds its way through the soil, many of the hydrogen ions (H^+) are exchanged with other cations, such as potassium and sodium, through ligand exchange. During wet years with a lot of rain, the importance of the catchment area is less pronounced than during dry years, and the water that reaches the lake has a chemical composition similar to rainwater. Moreover, the larger the catchment area is, the longer it takes for the water to reach the lake, and the more pronounced is the impact of the catchment area on lake water chemistry. Similarly, a lake in a large catchment area with intense agricultural activity is likely to have high nutrient levels and productivity. On the other hand, a catchment area with mainly coniferous forests on granite rock and poor soils is likely to harbour a lake with low nutrient concentration, high humic content, and probably low pH, since conifers generally create acidic soil. Furthermore, the smaller a catchment area is, the more likely it is that the lake has a low concentration of nutrients, since the rainwater has a short distance to gather nutrients before it reaches the lake. Hence, knowledge of the catchment area may allow for predictions about several characteristics of the lake. Moreover, the features of the catchment area to a great extent determine the abiotic frame that is presented to the organisms in the lake. Although knowledge about conditions in the catchment area cannot allow for more than rough guesses about the features of the lake or pond receiving water, a glance at a map, where land use and size of the catchment area are shown, may give valuable information about what to expect before a visit to a lake. One of the factors that is strongly affected by the catchment area, and which in turn affects the light climate of the waterbody, is the colour of the water.

Water colour

The colour of the water in a lake or pond is mainly derived from organic material, which are remains from dead plants and animals at different stages of decomposition. Of specific interest are the humic substances, which give bogs, as well as many streams, lakes, and ponds the colour of weak tea, or, in extreme cases, more like cognac, which has given them their German nickname *Cognac Seen* (brandy lakes). **Humic substances** are large molecules which are difficult for the microbial community to degrade, mainly owing to the phenols that are included in the molecule. This means that humic substances are long-lived and therefore accumulate in aquatic systems. Except for the colour, humic substances lead to lower transparency of the water, as well as to lower pH, and often to lower oxygen concentrations; features that slow down the metabolism and thereby productivity of the lake. Some of the humic substances are produced by organisms that have lived in the lake (**autochthonous** material), but the major portion generally stems from the catchment area (**allochthonous** material). Lakes and ponds in catchment areas dominated by coniferous forests are generally brown in colour because of the slow degradation of coniferous tree litter. Similarly, in areas where bogs are common the lakes and ponds can also be expected to have a brownish colour. Although low pH, oxygen, and light penetration suggest that humic lakes are low productive, the input of allochthonous material from the catchment is a potential energy source for organisms. Actually, despite that humic substances may be difficult for microorganisms to degrade, a considerable part of these substances are metabolized and transferred further up the food chain; that is, fuelling the whole ecosystem (Tranvik 1988). The discovery of such processes has increased the interest in studying humic, brown-water lakes.

The most widespread method to assess the colour of a specific waterbody is based on platinum units (Pt). This method uses a solution of potassium hexachloroplatinate (K_2PtCl_2) as a standard, which is then visually compared with the lake sample. A very clear polar lake may have a Pt value of zero, whereas bog water often has values above 300 mg Pt L^{-1}.

Carbon

Carbon enters the lake either as carbon dioxide from the air and is fixed by photosynthesizing organisms (autochthonous carbon), or from degradation of dead terrestrial organisms (allochthonous carbon). Carbon may also enter as bicarbonate through ground and surface water from the catchment area. In high productivity lakes, where the rate of photosynthesis is high, the major portion of carbon is autochthonous, whereas the allochthonous portion is high in humic lakes. Indeed, one of the more widely used lake typologies is based on where the carbon is produced and how much. Three types of lakes are identified: (1) **oligotrophic** lakes (from Greek: *oligotrophus* = low nutritious) where productivity is low and mainly based on carbon assimilated within the lake; (2) **eutrophic** lakes (from Greek: *eutrophus* = high nutritious) where the production is high, and also mainly based on internally assimilated carbon; and (3) **dystrophic** lakes (from Greek: *dystrophus* = malnutritious; see above) where production is mainly based on allochthonous carbon. Dystrophic lakes generally have low primary productivity and are often acidic due to high amounts of humic and other acidic substances entering the lake from the catchment area.

Carbon may be bound in living or dead organisms, or it may be dissolved in the water (dissolved organic carbon, DOC). The total amount of DOC in a lake is very large, although about 90% is recalcitrant (i.e. not easily available to organisms). Only about 10% is directly available as a resource to bacteria. Recently, the available portion of DOC has been shown to increase as a result of photochemical degradation (Lindell *et al.* 1995), that is, light energy splits large carbon molecules, even humic substances, into smaller molecules easier to degrade by bacteria and protozoa. UV radiation is especially important in this degradation. Since carbon cycling is fundamental in total lake metabolism, the effects of a global increase in UV radiation may thus have profound and unexpected effects on aquatic ecosystems (see also Chapter 6, *Biodiversity and environmental threats*).

Adaptations to obtain carbon

In plants, carbon is taken up as carbon dioxide (CO_2) or as bicarbonate (HCO_3^-) and used in photosynthesis. The relatively low concentrations of carbon dioxide in water compared to air may limit photosynthesis in submerged macrophytes. Another problem is that diffusion rate of CO_2 is about 10 000 times slower in water than in air. As submersed plants consume CO_2, microzones with low CO_2 concentration are created around the plant due to the slow diffusion rate through water. This may lead to a considerable reduction in photosynthesis rate. In dense stands of submersed macrophytes, the concentration of free CO_2 may thus become very low during daytime, whereas night-time concentrations are high due to respiration by both plants and animals. Unfortunately for the plant, photosynthesis needs light to fix carbon and therefore carbon dioxide cannot be taken up during the night by most aquatic plants. However, some plants are able to use **crassulacean acid metabolism (CAM)**, a subsidiary system to photosynthesis allowing night-time uptake of carbon dioxide which is transformed to malic acid and stored in vacuoles. During daytime, when the CO_2 level in the water is low due to plant uptake, the CAM plants convert the accumulated malic acid to CO_2, which is then used in photosynthesis. CAM metabolism is an adaptation to withstand low concentrations of carbon dioxide and may provide up to 50% of the total carbon requirement. CAM species, including genera such as *Isoetes* and *Littorella* (Fig. 3.4), are found mainly in lakes of low productivity where carbon levels are low. Plants that use CAM metabolism are only able to use free carbon dioxide and cannot utilize bicarbonate (HCO_3^-) as a carbon source as do many other aquatic macrophytes (e.g. *Myriophyllum* sp.). The disadvantage of using HCO_3^- instead of CO_2 is that it involves an active uptake process and is therefore energy costly, and the affinity of bicarbonate is lower than for carbon dioxide.

Some submersed aquatic plants (e.g. *Hippuris* sp. and *Ranunculus* sp.), have both submersed and emergent leaves (**heterophylly**) allowing CO_2 uptake both in water and from the atmosphere, where diffusion rates and concentration of CO_2 are higher than in the water. Due to adaptations, such as *CAM metabolism, heterophylly*, and the use of HCO_3^- *as a carbon source*, the problem of carbon limitation in aquatic plants may be of minor importance. Instead, other factors, such as light, nutrients, and oxygen supply to the root system, usually become limiting before carbon.

Mixotrophy

A majority of the organisms on earth either rely solely on photosynthesis (autotrophy) or solely on assimilation of organic substances (heterotrophy) to meet their requirements for energy and carbon. However, some organisms are able to combine autotrophy and heterotrophy and are, accordingly, called **mixotrophic**. The benefits with this mixed feeding strategy is obvious—they can be 'plants' when it is beneficial to use solar energy and they become 'animals' when that feeding strategy is more efficient (Fig. 2.12). There are, however, significant energetic costs in having two different feeding systems; costs that have to be outweighed by the benefits (Rothhaupt 1996), leading to that a mixotrophic feeding strategy is not always beneficial compared to relying entirely on either phagotrophy (feeding on other organisms) or autotrophy (using light as energy source) (Fig. 2.12).

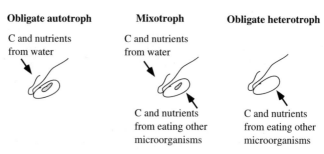

Fig. 2.12   Different feeding systems among, for example, flagellates: obligate autotrophs using chlorophyll (illustrated as a grey dot in the cells) for light acquisition and absorbing carbon and nutrients from the water, and obligate heterotrophs eating other microorganisms. The mixotrophs use a combination of these feeding systems.

Mixotrophy is most common among single-celled organisms, such as flagellated algae and ciliates, but have also been found among sponges and rotifers (Tittel *et al.* 2003). In freshwaters, mixotrophic organisms are generally most common in oligotrophic and humic lakes, which is due to poor light conditions for photosynthesis, but also due to the often low concentrations of inorganic nutrients in such lakes (Jones 2000). Moreover, the high concentration of dissolved organic matter in humic lakes leads to high abundances of bacteria; organisms suitable as food packages for mixotrophic organisms. Hence, carbon and energy can be obtained either by photosynthesis or by feeding on other microorganisms, which can also provide other nutrients, such as nitrogen, phosphorus, and vitamins (Fig. 2.12). Therefore, it is not surprising that mixotrophs are generally most successful in humic lakes, whereas obligate phototrophs and heterotrophs do better in other types of water.

Another situation where mixotrophy may be beneficial is where both bacteria and algae are limited by a certain nutrient, such as phosphorus or nitrogen. Since bacteria are generally competitively superior to the algae with respect to nutrient uptake, the algae will have problems. This offers an opportunity for mixotrophs since they can feed directly on bacteria, thereby both reducing the abundance of a competitor and taking the nutrients they have assimilated, and simultaneously utilize photosynthesis (Jansson *et al.* 1996). Although the phenomenon of mixotrophy has been known for long, it was not until the mid-1980s its importance was documented. Then it was demonstrated that the common algal genera *Dinobryon* (Fig. 3.6), besides using photosynthesis, also fed vigorously on bacteria and reduced bacterial density with 30% (Bird and Kalff 1986).

Autotrophic versus heterotrophic systems

Input of carbon from terrestrial sources (allochthonous carbon), such as leaves, debris, and humic compounds, provides a resource that is metabolized

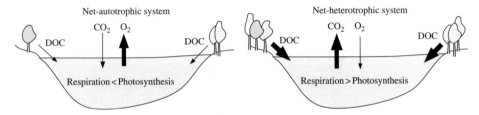

Fig. 2.13   In net-autotrophic systems photosynthesis rate exceeds respiration rate and oxygen is exported to the atmosphere. In net-heterotrophic systems organisms are subsidized with carbon (DOC, POC) from the catchment and the respiration rate becomes higher than photosynthesis rate leading to that carbon dioxide is exported to the atmosphere.

by bacteria and passed forward through the food chain and incorporated also in other organisms. Hence, terrestrial input of DOC and particulate organic carbon (POC) can enhance organism abundances beyond levels supported by photosynthesis in the lake; the system has become net-heterotrophic; that is, the respiration exceeds photosynthesis (Fig. 2.13). Although a system may be mainly net-autotrophic or mainly net-heterotrophic the temporal variation in both respiration and photosynthesis is considerable, leading to that the Net Ecosystem Production (NEP) in a specific lake may be higher than zero (photosynthesis > respiration) during periods of high photosynthesis, for example, during summers and lower during other periods. Moreover, changes in the food web composition may also affect the NEP which was elegantly shown experimentally in some North American lakes (Cole *et al.* 2000). The NEP was positive (photosynthesis > respiration) only at high nutrient loading combined with high predation pressure on zooplankton from fish. The reduction of grazing zooplankters combined with high nutrient availability lead to higher algal biomass, and thereby photosynthesis. Hence at these conditions, oxygen production by photosynthesis increased and the system moved from being net-heterotrophic to net-autotrophic.

pH

pH, which is a measure of the acidity of a solution, is another important abiotic factor. It is defined as the logarithm of the reciprocal of the activity of free hydrogen ions ($\log[1/H^+]$), which means that a change of one unit in pH corresponds to a 10-fold change in the activity of hydrogen ions. Lakes and ponds show regional differences in pH due to differences in geology and hydrology of the catchment area, input of acidifying substances, and productivity of the system, but the pH in the majority of lakes on earth is between 6 and 9 (Fig. 2.14). However, in volcanic regions, freshwaters

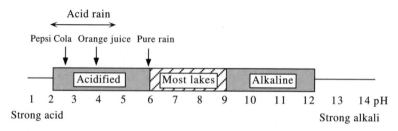

Fig. 2.14   Range of pH in lakes in relation to strong acid and alkali liquids common in everyday life. 'Acidified' includes lakes with extremely low pH in volcanic regions receiving high amounts of sulphuric acid, and naturally acidic bog lakes, as well as lakes acidified by human activities. Most lakes on earth have pH values between 6 and 9. Note that this is a schematic picture with numerous exceptions.

may have a pH as low as 2 owing to input of strong mineral acids, mainly sulphuric acid. Bog lakes may have a pH of 4 or lower, mainly because of the active cation exchange in the cell walls of *Sphagnum* mosses, a plant often dominant in these systems. This exchange releases hydrogen ions to the water, thus lowering the pH. Hence, in this case the dominating organism, *Sphagnum*, may considerably alter the abiotic frame, thus affecting the success of other organisms colonizing the system. Pure rain, not affected by acidification, has a pH of around 6, whereas acid rain may reach pH values as low as 2 (Fig. 2.14). In such cases, rain may have a considerable impact on the lake ecosystem (see Chapter 6, *Biodiversity and environmental threats*).

pH, alkalinity, and the carbon dioxide–bicarbonate complex

pH is also strongly related to equilibrium processes of the carbon dioxide–bicarbonate system, including free carbon dioxide (CO_2), carbonic acid (H_2CO_3), bicarbonate ions (HCO_3^-), and carbonate ions (CO_3^{2-}). Atmospheric carbon dioxide is very soluble in water and when dissolved it is in equilibrium with the weak acid carbonic acid H_2CO_3 (Fig. 2.15). At increasing pH, carbonic acid dissociates into a hydrogen ion (H^+) and a bicarbonate ion (HCO_3^-), which in turn dissociates into another H^+ ion and a carbonate ion (Fig. 2.15). Photosynthesis and respiration are the major biological processes affecting pH by changing the amount of CO_2 in the water. The photosynthesis of green plants uses solar radiation and carbon dioxide to produce sugar and oxygen. The consumption of CO_2 alters the equilibrium, causing an uptake of hydrogen ions and thereby an increase in pH (Fig. 2.15). In calcium-rich regions, photosynthesis may cause precipitation of calcium carbonate ($CaCO_3$), as a result of CO_2 consumption by the plants. The result of this reaction can be seen as white or grey encrustations covering leaf surfaces of macrophytes. As organisms respire, CO_2 is produced, pushing the equilibrium reaction in the opposite direction releasing H^+ ions and thereby reducing pH.

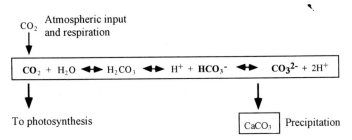

Fig. 2.15   The pH–carbon dioxide–bicarbonate system. When carbon dioxide (CO_2) is taken up by photosynthesis, free H^+ ions are associated with HCO_3^- and CO_3^{2-}, leading to fewer free H^+ and thereby a higher pH. At very high photosynthesis, pH may be raised to the point where the carbonate ions are precipitated as calcium carbonate. On the other hand, respiration adds carbon dioxide to the system which releases H^+ ions, leading to lower pH. Figure based on Reynolds (1984).

In regions with a bedrock rich in carbonates, such as $CaCO_3$, the weathering of carbonates makes the water well buffered against acid substances. Such waters are said to have high **alkalinity**, or a high acid neutralizing capacity. If the alkalinity is zero, the pH drops even at small additions of acid, whereas if the alkalinity is higher, the pH does not decrease proportionally to the acid (H^+) addition. Thus, alkalinity is a measure of how sensitive a lake or pond is to acidification. It also means that alkalinity is a better estimate than pH of a lake's or pond's status with respect to acidification, since the capacity of withstanding further acid additions is more stable than the instantaneous concentration of H^+ ions, which may vary on a temporal scale of hours due to, for example, photosynthesis and respiration (see above). Alkalinity is generally expressed as milliequivalents per litre (meq l^{-1}). Lakes with an alkalinity above 0.5 meq l^{-1} are categorized as having very good buffering capacity and thereby a low risk of becoming acidified. The other extreme is waters with alkalinities below 0.01 meq l^{-1}, which have very low, or no, buffering capacity and are therefore extremely vulnerable to acidification. As you may have understood from the text above, there is a rough empirical relationship between pH and alkalinity which may be expressed as:

$$pH = 7.3 + 0.82 \log(\text{alkalinity})$$

It should, however, be noted that this relationship is only a rule of thumb which may vary between regions and lake types. Moreover, the relationship is only valid at pH above 5.4 and alkalinities above 0.005 meq l^{-1}. As noted above, pH is strongly affected by biological processes such as photosynthesis and respiration. Although alkalinity shows less fluctuation over time than pH within a specific lake or pond, it is not unaffected by biota. For example, when bacteria oxidize inorganic sulphur or nitrogen compounds, H^+, SO_4^{2-}, and NO_3^- are formed, leading to reduced alkalinity. On the other hand, plant uptake of NO_3^- and microbial reduction of SO_4^{2-} and NO_3^-, leads to increased alkalinity. Such microbial redox reactions may have profound effects on alkalinity and pH when waterlogged sediments are aerated, for example, when a pond or a wetland dries out. The aeration of previously anoxic sediments leads to oxidation of reduced sulphur minerals and thereby a release of H^+ ions. When water comes back and again covers the sediment, the pH may decrease considerably with, sometimes, catastrophic effects on the organisms.

Reduction in pH negatively affects the reproduction of many organisms, including crayfish, *Daphnia*, molluscs, insects, and many fish species, leading to a reduction in abundance of these organisms. The phytoplankton community switches towards dominance of large flagellated algae, such as *Peridinium* or *Gonyostomum*, whereas the zooplankton assemblage becomes dominated by large copepods (e.g. *Eudiaptomus*) instead of daphnids. Obviously, some organisms are more tolerant to low pH than others, although the specific physiological adaptations to withstand low pH are largely unknown. We will return to community effects of acidification in Chapter 6 (*Biodiversity and environmental threats*).

Nutrients

Nutrients are chemical elements that organisms require as 'building blocks' in their cells. The quantity of nutrients in lake water is mainly determined by bedrock type, vegetation cover, size, and human activities in the catchment area. Since nutrients are dissolved in the water, animals cannot absorb them directly. Instead, bacteria, algae, and other primary producers, able to absorb dissolved molecules and concentrate nutrients, can be considered as suitable nutrient packages for herbivores, such as crustaceans. The latter are then used as food packages for larger predatory organisms, such as fish. In this way nutrients are transported upwards through the food web. Most of the ions required for plant and animal growth are supplied from soils and rocks in the catchment area. The majority of these ions are available in much higher concentrations than required by the organisms. As illustrated by the low availability–demand ratio (Table 2.1), and as we will see later, phosphorus is the element most likely to be growth-limiting for algae in freshwaters.

Measuring chemical relations—ecological stoichiometry

Everything, both abiotic structures as sand grains and clay particles, and biotic components as virus and fish, is composed of chemical elements, such as nitrogen (N), phosphorus (P), and carbon (C). These elements are constantly transported through the ecosystem by processes such as weathering

Table 2.1 Relative availability and demand by plants and algae of different elements in relation to phosphorus, which has been set to a value of 1 (partly based on Moss 1980). The main function of each element is also indicated. If the ratio of availability to demand is greater than 1, the requirement of the element is more likely to be met by supply than is the case for phosphorus

Element	Availability	Demand	Availability/ Demand	Function
Na	32	0.5	64	Cell membrane
Mg	22	1.4	16	Chlorophyll, energy transfer
Si	268	0.7	383	Cell wall (diatoms)
P	1	1	1	DNA, RNA, ATP, enzymes
K	20	6	3	Enzyme activator
Ca	40	8	5	Cell membrane
Mn	0.9	0.3	3	Photosynthesis, enzymes
Fe	54	0.06	900	Enzymes
Co	0.02	0.0002	100	Vitamin B_{12}
Cu	0.05	0.006	8	Enzymes
Zn	0.07	0.04	2	Enzyme activator
Mo	0.001	0.0004	3	Enzymes

of rocks, growth, and death of animals and plants, as well as their assimilation, excretion, and decomposition. The study of relations, for example ratios, among chemical elements is called stoichiometry (from the Greek word *stoicheion* which means element); a word borrowed from chemistry. The expression *ecological stoichiometry* has, however, become increasingly important in ecology, especially in aquatic ecology (Elser and Hassett 1994; Sterner and Elser 2002), which is partly due to findings that can be simplified in the phrase: 'you are what you eat'. This means that the chemical composition of your body reflects the chemical composition of your food. If a consumer's chemical composition should perfectly mirror that of its food, a plot on food and consumer stoichiometry, for example, the ratio between nitrogen and phosphorus (N : P), results in a straight line with the slope 1 (Fig. 2.16). Although the statement 'you are what you eat' is appealing and widely used also with respect to humans, for example, in advertisements for health food, the reality is more complex than so. Many organisms discriminate among elements, that is, they are not completely passive consumers assimilating all chemicals in their food, but strive to keep the chemical composition of their body constant irrespective of the chemical composition of their food; they are **homeostatic** (Fig. 2.16). A strictly homeostatic organism keeps the chemical composition of its body constant irrespective of what the chemical composition of the food is. Hence, a plot of the stoichiometry of the food and of a homeostatic consumer will show a line parallel to the food axis (Fig. 2.16). Of course, there exist all possible compromises between none and strict homeostasis. Although, stoichiometry may not always be as simple as—you are what you eat—the complexity does not make stoichiometry less interesting or relevant with respect to ecology. Instead it may

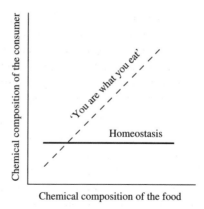

Fig. 2.16 Generalized relationships between the chemical composition of food and its consumer. The dashed line illustrates how the body chemistry of the consumer mirrors that of its food. The solid line shows the other extreme when the consumer keeps its chemical composition constant irrespective of changes in the chemistry of the food; that is, strict homeostasis.

explain many patterns and provide new insights with respect to natural systems. One such insight is that maintaining homeostasis requires energy; that is, there is a cost in keeping the internal stoichiometry different from that of the food. Assuming that spending energy affects the performance of the individual organism, for example the number of offsprings, homeostasis and stoichiometry may be worth considering when studying everything from organism interactions to ecosystems.

The Redfield ratio

Although stoichiometry is applicable to any element, most studies are focused on C, N, and P, mainly because these are major constituents of organisms, and, at least with respect to N and P, often limiting for growth. Another reason making studies of the relation among C, N, and P interesting was the finding by the oceanographer Alfred Redfield (1890–1983) that the relative amount of these elements in planktonic organisms, as well as in the surrounding water, was 106 : 16 : 1 (by atoms). This means that for every phosphorus atom there are 16 nitrogen and 106 carbon atoms. The Redfield ratio has become very influential and is one of the few 'laws' or at least rule of thumb, in limnology and oceanography.

Phosphorus

Phosphorus is essential for all organisms since it is used in fundamental processes such as storage and transfer of genetic information (DNA and RNA), cell metabolism (various enzymes), and in the energy system of the cells (**adenosine triphosphate, ATP**).

Phosphorus is taken up as phosphate (PO_4^{3-}), which is the only inorganic fraction of phosphorus of importance for organisms. Most phosphorus in the water, usually above 80%, is included in the organic phosphorus fraction (i.e. incorporated in organisms). The sum of all phosphorus fractions, including inorganic and organic forms, is called total phosphorus (TP) and is widely used as an estimate of lake fertility.

Phosphorus enters the lake water from the *catchment area* and via *sediment release*, but also via *atmospheric deposition*. Most natural lakes (not affected by man) have phosphorus concentrations of between 1 and 100 µg l^{-1} (Fig. 2.17). Due to human impact during the last 50 years, a major proportion of lakes close to urban regions have considerably higher phosphorus concentrations than undisturbed lakes. Since phosphorus is generally limiting for the growth of primary producers in freshwaters, it is also the main determinant for production. Phosphorus concentration is easier to measure than carbon content and production and lakes are therefore often classified as 'oligotrophic' and 'eutrophic' based on phosphorus concentration. Hence, lakes with low phosphorus concentrations (5–10 µg total phosphorus l^{-1}), and thereby low productivity, are categorized as 'oligotrophic'. Lakes with between 10 and 30 µg phosphorus l^{-1} are categorized as 'mesotrophic', whereas lakes with high phosphorus concentrations (30–100 µg l^{-1}) are 'eutrophic'. Below 5 µg l^{-1} and above 100 µg l^{-1} total phosphorus, lakes are 'ultraoligotrophic' and 'hypereutrophic', respectively.

Sediments are generally richer in phosphorus than lake water and many factors can affect the transport of phosphorus between sediment and water. One such factor is pH, which affects the exchange of phosphorus between sediment and water. At pH values below 8, phosphate bindings to metals are strong, whereas at higher pH values, hydroxide ions (OH^-) are exchanged with the phosphate, which then becomes soluble in water. This may be an important part of the input of phosphorus to the water of eutrophic lakes with high primary production: primary production increases the pH of the water, and thereby promotes phosphorus release, which in turn promotes even higher algal production!

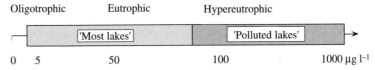

Fig. 2.17   Range of concentration of total phosphorus in lakes (µg l^{-1}). 'Most lakes' refers to lakes not affected by human activities, ranging from hyperoligotrophic melt water lakes in polar regions (total phosphorus around 5 µg l^{-1}) to hypereutrophic with large catchment areas. 'Polluted lakes' refers to lakes receiving nutrient input from human discharge water or from high concentrations of animals, such as bird colonies. Note that this is a schematic diagram with numerous exceptions.

Nitrogen

Nitrogen mainly occurs in amino acids and proteins of organisms and may enter the lake by *precipitation*, by *fixation* of atmospheric nitrogen (N_2), or by input from surface and groundwater *drainage*. Nitrogen concentrations in lakes vary widely from about 100 $\mu g\, l^{-1}$ to over 6000 $\mu g\, l^{-1}$ (Fig. 2.18). Since nitrogen is generally not the main limiting nutrient for organisms in freshwaters, its concentration in water is less strongly connected to lake trophic state than phosphorus, which means that an oligotrophic lake may not necessarily have low nitrogen concentrations. In many polluted lakes, however, the phosphorus concentration may become very high, leading to nitrogen being the main limiting nutrient for algal growth. A significant amount of nitrogen in a lake is bound in organisms (organic N), but it also occurs as molecular N_2, nitrate (NO_3^-), nitrite (NO_2^-), and in the reduced form ammonium (NH_4^+). Nitrate and nitrite have to be reduced to ammonium before they can be assimilated in the cell, making ammonium the most energetically favourable nitrogen source for cell uptake. The distribution of nitrogen in its oxidized (nitrate) and reduced (ammonium) form differs considerably between eutrophic and oligotrophic lakes. In oligotrophic lakes, ammonium occurs only in low amounts and shows little variation with depth due to the presence of oxygen throughout the water column. In eutrophic lakes, however, processes are more complicated owing to a high outflow of ammonium from the sediment as a result of bacterial decomposition of organic material at low oxygen concentrations (Fig. 2.19). The concentration of nitrate generally follows the oxygen curve. In eutrophic lakes, where the oxygen concentration is low in deeper areas, most of the nitrate is reduced to nitrogen gas (N_2).

Nitrogen fixation

Although primary producers are generally phosphorus-limited in freshwaters, they may occasionally become limited by nitrogen in eutrophic lakes and ponds, where phosphorus concentrations are high. In such situations, certain cyanobacteria and bacteria are able to fix atmospheric nitrogen, an adaptation that is exclusive for these groups. Hence, during periods when nitrogen is limiting for plant growth, cyanobacteria are likely to dominate the algal community. Nitrogen fixation by cyanobacteria may in some cases contribute up to 50% of the annual nitrogen input to a lake. The fixation of

Fig. 2.18   Range of concentration of total nitrogen in lakes (summer values; $\mu g\, l^{-1}$). 'Most lakes' refers to lakes not affected by human activities. 'Polluted lakes' refers to lakes receiving nutrient input from human discharge water or from high concentrations of animals, such as bird colonies. Note that this is a schematic diagram with numerous exceptions.

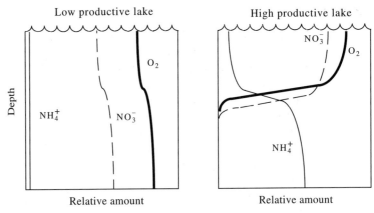

Fig. 2.19 Vertical distribution of oxygen (O_2), nitrate (NO_3^-), and ammonia (NH_4^+) in stratified lakes with low productivity (oligotrophic) and high productivity (eutrophic). The nitrate curve mirrors the oxygen concentration, whereas the relative amount of ammonia increases at low oxygen concentrations.

nitrogen occurs mainly in **heterocytes** (Fig. 3.6), where oxygen is absent, a necessary requirement for the catalyst enzyme, nitrogenase. Fixation may also occur within bundles of algae, such as in the cyanobacteria *Aphanizomenon* (Fig. 3.7), where oxygen concentration is low. Algal nitrogen fixation occurs in light since it requires energy in the form of ATP which is derived from photosynthesis. Many cyanobacteria are capable of nitrogen fixation, whereas only a few groups of heterotrophic bacteria fix nitrogen (*Azotobacter* and *Clostridium*).

Nitrification and denitrification

Many bacteria, and to a minor extent some fungi, are involved in the transformation of ammonium to nitrate (NO_3^-, **nitrification**) and in the transformation of nitrate to molecular nitrogen (N_2, **denitrification**). Nitrification requires oxygen and the reaction can be summarized as:

$$NH_4^+ + 2O_2 \rightarrow NO_3^- + H_2O + 2H^+ \text{ (Nitrification)}$$

In the nitrification process the bacteria oxidizes ammonium (NH_4^+) to nitrate (NO_3^-), a reaction that provides them with energy.

Denitrification, on the other hand, occurs intensely under anaerobic conditions, such as in anoxic sediments and in anaerobic microzones, and may be defined as the reduction of oxidized nitrogen during the oxidation of organic matter (decomposition). Hence, denitrifying bacteria gain energy by oxidizing organic carbon to carbon dioxide (CO_2) and carbonate (CO_3^{2-}) by using oxygen from nitrate. Simultaneously, the released electrons from the oxidation are transferred to nitrogen which is in steps reduced to gas (N_2). In addition to energy, denitrifying bacteria gain carbon which they use to build-up new biomass. The denitrification reaction can be summarized as follows:

$$5C_6H_{12}O_6 + 24NO_3^- \rightarrow 12N_2 + 18CO_2 + 12CO_3^{2-} + 30H_2O$$
(Denitrification)

The molecular nitrogen (N_2) is lost to the atmosphere if not used in nitrogen fixation. Note that nitrification occurs in the presence of oxygen (aerobic environment), whereas denitrification is an anaerobic process. This means that these processes are performed by different types of bacteria. Nitrification produces nitrate which is used in denitrification, and thus, the processes are often coupled. Hence, nitrification and denitrification are often intense in aerobic–anaerobic interfaces, such as at the sediment surface and in chironomid burrows in the sediment, as well as around roots and rhizomes of macrophytes.

Micronutrients

The supply of the micronutrients cobalt (Co), copper (Cu), zinc (Zn), and molybdenum (Mo) from surrounding rocks and soils is very low, but since the demand for these elements is even lower, they are rarely limiting for plant growth (Table 2.1). However, high concentrations of these elements are toxic to both plants and animals. One example is copper sulfate, which has been used successfully to eliminate algal blooms in lakes. Micronutrients are mainly used as ingredients in specific enzymes, functioning as catalysts for chemical reactions in the cell.

Potassium (K) and sodium (Na) are important components in cell membrane processes, whereas silica (Si) is used in large quantities by diatoms and chrysophytes for the construction of cell walls and silica scales, respectively. Silica may be depleted during diatom blooms, which results in a rapid decline in the abundance of diatoms.

Calcium (Ca) is used mainly for the formation of the skeleton of vertebrate animals, but has never been reported to limit growth in higher aquatic animals. However, molluscs use high amounts of calcium in their shells. In areas with low calcium supply from weathering rocks, molluscs are unable to construct their shells, resulting in this group of animals being excluded from the fauna in calcium-poor areas. This is an example of how an axis of the abiotic frame (in this case the calcium concentration) can exclude an organism with a specific adaptation (calcium-rich shells to reduce predation), which may otherwise have been highly competitive.

The cycling of nutrients

All elements, micro- as well as macronutrients, are transported within and between lakes and their environment, transformed between different

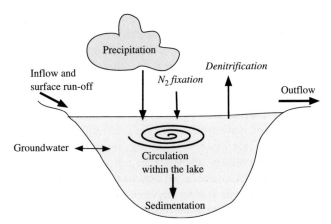

Fig. 2.20   General pathways for transport of chemical substances to, from and within a lake. With respect to nitrogen (N) two specific pathways are indicated: nitrogen fixation from the atmosphere and denitrification.

chemical compositions, and translocated within and between organisms. Although elements differ somewhat in how they are moving around in the ecosystem, most of them follow a general *nutrient cycle*. From a lake's point of view, elements can enter either through a creek or river, via surface run-off from the surrounding land, via rainfall, or through the groundwater (Fig. 2.20). Once in the lake, elements may be taken up by organisms and transported through the food web (see Chapter 5), sinking down to the sediment, entering the groundwater, or following a creek or a river out of the lake (Fig. 2.20). Although much can be said about variation in cycling among elements, this is the general way they are transported within a lake and exchanged between the lake and the surroundings and we will, with one exception, not go further into detail here. The exception is nitrogen which, in contrast to most other elements, has a quantitatively important atmospheric phase; that is, about 80% of the atmosphere is N_2 gas. As we have seen in the previous sections, some microorganisms are able to fix nitrogen from the atmosphere (nitrogen fixation of N_2) and others are able to transform nitrate to N_2 through denitrification. So, with respect to nitrogen we have to add to the general cycle one arrow leading out from the lake (denitrification) and one leading into it (nitrogen fixation) (Fig. 2.20).

Limiting nutrients and adaptations to low nutrient availability

Carbon (C), nitrogen (N), and phosphorus (P) are not particularly abundant on earth, but they are the main constituents of molecules and processes in organisms, such as in proteins, energy transfer, and support tissue. Generally, rapidly growing algae have uptake ratios of carbon, nitrogen, and phosphorus in the proportions 106 C : 16 N : 1 P (by atoms;

the Redfield ratio). The ratio between nitrogen and phosphorus (N : P) in the surrounding water is often used to assess which of these nutrients that is limiting for phytoplankton. If the N : P in the water is higher than the Redfield ratio (N : P > 16) the algae are most probably phosphorus limited, since nitrogen is then in excess compared to phosphorus. The opposite is true when N : P is lower than 16. The relatively *low supply rate* of phosphorus and nitrogen in combination with the *high demand* are major reasons why they are often limiting for growth among organisms. Moreover, since all dead organisms end up at the bottom of a lake, the sediments are often rich in nitrogen and phosphorus, and therefore not instantly available to organisms in the water column. In freshwaters, phosphorus is often the major limiting nutrient, although nitrogen may under certain conditions become limiting. One reason for the phosphorus limitation is that run-off water from unpolluted catchments generally has N : P ratios as high as 30, suggesting that *P is provided in lower rates than N*. Since freshwaters are much smaller and shallower than oceans they are more directly affected by the impact from surrounding land, rivers, and streams (the catchment), which may be a reason why organisms in freshwaters are generally P limited. Another reason for this difference between fresh and marine waters may be that the latter have high concentrations of sulphur (S) and low concentrations of iron (Fe). This leads to more extensive sequestering of iron by sulfides in marine waters; that is, less phosphate molecules form precipitates with iron, and instead remain in solution where they are available to organisms (Caraco *et al.* 1989; Blomqvist *et al.* 2004). In freshwaters, on the other hand, iron is more common and forms chemical complexes with dissolved phosphorus, which precipitate and become less available to algae and bacteria (see below: 'Remineralization of nutrients').

A whole lake experiment convincingly showing that phosphorus is growth limiting in freshwaters was performed by Schindler in 1974. Then a lake was divided into two basins, and phosphorus, nitrogen, and carbon were added to one of the basins while only nitrogen and carbon were added to the other. An algal bloom occurred in the first basin, whereas the water stayed clear in the latter (Schindler 1974). This simple whole lake experiment was a milestone in applied limnology since it demonstrated that phosphorus-rich detergents were the likely factor behind eutrophication of lakes (see also Chapter 6, *Biodiversity and environmental threats*). Moreover, this experiment also illustrates that if a lake is used as a recipient for human wastewater, for example, from a sewage treatment plant, the ratio between N and P often decreases to below 10 since animal, including human, faecals are very rich in P. This means that algae and other plants in polluted lakes often show N-, instead of P-limited growth.

In order to obtain as much of this limiting resource as possible, freshwater plants have several adaptations: they may have a small *size* and thereby increase the uptake rate; they can store phosphorus when availability is high (*luxury uptake*); and they can *invade habitats* with high concentrations of phosphorus, such as the sediment.

Size

Uptake of phosphorus by algae and microorganisms is usually extremely rapid, illustrated by an experiment where 50% of the added radioactive phosphate was taken up 30 s after addition (Lean 1973). However, uptake rates vary among algal species and may affect species composition in algal communities. Generally, small size is favoured by frequent nutrient pulses of low concentrations, whereas less frequent pulses with higher concentrations of nutrients favour larger forms (Turpin 1991). Hence, the patchiness in nutrient supply may force the algal community towards large or small forms. Since rain brings nutrients from the catchment area to the lake as a short pulse, the frequency and duration of rainfall (i.e. the climate), may have a considerable impact on the size distribution of phytoplankton in a lake!

Small size (i.e. having a high surface area to volume ratio, SA : V) increases the uptake surface area in relation to the 'consuming' volume. Transport of nutrients and excretion products within the cell must rely on diffusion and transport at the molecular level, and thus cells that aim at optimizing nutrient uptake rates should be small, resulting in a favourable ratio of absorptive surface to volume ratio. Despite this, cell and colony length in algae have a considerable range (from 4 μm *Synechococcus* to 200 μm *Microcystis*) with SA : V ratios of 1.94 and 0.03, respectively. Hence, it may not come as a surprise that other selection pressures favour large cells, including higher resistance to grazing and, as noted above, that larger cells are favoured when nutrient pulses are higher in concentration and less frequent.

Luxury uptake

An interesting adaptation to secure nutrient availability is excess ('luxury') uptake during periods when a nutrient that is usually limiting occurs in high concentrations. With respect to phosphorus, many algal groups are able to produce polyphosphates which are stored in small vacuoles. When needed, these can be broken down into phosphate molecules by enzymes, and used in cell metabolism.

Occupying a nutrient-rich habitat—the sediment surface

Organisms contain high amounts of nutrients, and since faecal pellets and dead organisms sink to the bottom, the sediment becomes enriched with nutrients. The amount of sinking organic material reaching the sediment is generally in proportion to the intensity of photosynthesis, and thus mirrors the lake's productivity. Hence, the sediment is often an important nutrient source, and algae associated with the sediment are rarely growth-limited by lack of phosphorus, as is often the case for organisms in the water column.

Remineralization of nutrients: effects of redox potential and pH   Decomposition of organic material occurs mainly in the upper millimetres of the sediment surface. As decomposition proceeds, nutrients will again be available for organisms in the water column. This recycling is governed by several mechanisms, including both abiotic and biotic processes. One of the main chemical processes affecting the flux of phosphorus at the sediment–water interface is the complex binding of phosphate with iron in the presence of oxygen, and the release of phosphorus during anoxic conditions (Boström *et al.* 1982). To understand this highly important oxidation–reduction reaction we have to introduce a few concepts from chemistry. A chemical process where electrons are lost is called *oxidation* and the opposite process, where electrons are gained, is called *reduction*. A reduction is always coupled with an oxidation and are together called a reduction–oxidation reaction, or a *redox reaction*. At equilibrium a redox reaction has a characteristic **redox potential** (E) which is measured against the standard zero point of hydrogen (H) and is therefore referred to as the E_H. The redox potential can easily be measured with a platinum (Pt) electrode and a reference electrode. When immersed in water and joined to a circuit the difference in potential between these two electrodes is measured with an ordinary voltmeter. A reducing solution has a tendency to donate electrons to the platinum electrode, whereas an oxidizing solution accepts electrons. In natural lake water or in sediments a lot of chemical reactions are going on simultaneously and redox reactions seldom or never reach equilibrium due to the continuously ongoing biological processes. Hence, what is shown on the voltmeter display is a rough estimate of the sum of all redox reactions going on, whereas little guidance is given about any specific reaction. Instead, redox potential measurements are useful to characterize a water or a sediment with respect to oxidative state, since a low E_H (below about $+100$ mV) is generally related to anaerobic (low oxygen concentrations) conditions. Accordingly, a high redox potential is related to well oxygenated environments. This mainly due to that dissolved oxygen gas (O_2) has a very high affinity for electrons (e^-). Hence, the oxygen concentration in the water or sediments can affect which chemical reactions that occur. An example of a chemical reaction of great importance for lake nutrient dynamics, and that is strongly affected by the oxygen concentration, is the redox reaction between the oxidized (Fe^{3+}) and the reduced (Fe^{2+}) forms of iron (Fe):

$$Fe^{2+} \leftrightarrow Fe^{3+} + e^-$$

At low oxygen concentrations (low redox potential), iron occurs as Fe^{2+} which is soluble in water. If the oxygen concentration increases (the redox potential increases), the chemical equilibrium of the redox reaction shown above is forced to the right and iron is transformed to its oxidized form (Fe^{3+}), which is not soluble in water but forms complexes with other molecules, precipitates, and sinks to the bottom of the lake. Phosphate (PO_4^{3-}), which is the main actor in the eutrophication process (see Chapter 6), has a high affinity for forming such precipitates. Iron is a common element in lakes (Table 2.1) and *if* oxygen is available the complexes of Fe^{3+} and PO_4^{3-} will precipitate and sink to the sediment, making phosphate less available for phytoplankton. However, as soon as the oxygen concentration declines, iron will transform to its reduced, soluble form (Fe^{2+}) and release its phosphate, again making it available for phytoplankton growth; a process termed **internal loading**. Hence, the flux of phosphorus between the sediment surface and the water is to a large extent determined by the oxygen availability. Internal loading of nutrients, for example phosphorus constitutes a major problem in lakes that have been used as recipients for polluted water. The sediment may function as a sink for nutrients during a period of years, but sooner or later the accumulated sediments will act as a nutrient source for the overlying water column. This situation causes problems when attempting to restore eutrophic waters, as we will see in Chapter 6 (*Biodiversity and environmental threats*).

In most lakes, not only polluted ones, changes in the redox potential will lead to phosphorus release to the water during stratified conditions when oxygen concentration is low, which for temperate, dimictic lakes occur during summer and winter. Since few organisms are able to survive in the hypolimnion during low oxygen concentration, the consumption of phosphorus will be low and phosphate will accumulate below the thermocline. When stratification is broken, during spring and autumn overturns, the accumulated phosphate will be available and rapidly consumed by organisms in the former epilimnion.

The considerable amount of organic material at the sediment surface leads to high abundances of bacteria using this material as substrate. During the mineralization process, bacteria use oxygen, that is, they are not only recycling phosphate, but are also indirectly making complex bound phosphorus available by reducing the oxygen concentration. In areas where light reaches the sediment, algae often cover its surface. In contrast to bacteria, algae *produce* oxygen, thereby creating an oxidized microzone, which will reduce phosphorus flow from the sediment to the water. Moreover, sediment-associated algae may consume considerable quantities of phosphorus which will be recirculated within the algal biolayer, further reducing the flux to the water. Hence, the metabolism of sediment-associated algae and that of bacteria affect the phosphorus dynamics at the sediment surface in opposite ways. Moreover, as we noted in the pH section above, phosphorus bound to metal ions in the sediment is released to the water by the increased pH induced by high algal photosynthesis. In highly productive (eutrophic) shallow lakes this pH increase caused by the high algal biomass during summer may, together with the redox reaction between Fe^{3+} and Fe^{2+}, be a process behind the internal loading of phosphorus.

Bioturbation

Particles, such as dead organisms, sink to the bottom and form new sediment. Since the sedimentation rate of organic material is generally higher during summer than during winter in temperate lakes, more or less distinct sediment layers are formed in the same manner as annual tree rings! These layers are generally visible in sediment cores taken from the anaerobic bottom layer where oxygen concentration is too low for benthic animals to live. However, in oxygenated sediments where animals can live, sediment layers are disturbed by mechanical mixing. This **bioturbation** is caused by foraging fish, worms, insect larvae, and mussels moving around and mixing sediment layers. Since the activity of aquatic animals is dependent on temperature, the bioturbation is most intense during summer.

Different animals occupy different parts of the sediment and the vertical, as well as horizontal patchiness in distribution, is tremendous. Worms, such as oligochaetes, and mussels may burrow, and thus affect the sediment layers more than 10 cm down into the sediment, whereas chironomids are most abundant in the top few centimetres. At the sediment surface of the **littoral** zone (Fig. 2.21), which is above the thermocline, the diel, as well as seasonal, variations in physical and chemical factors are significant, and the biodiversity of benthic animals is high. Hence, in the littoral zone, the bioturbation is intense and distinct sediment layers are seldom seen. However, in sediments situated below the thermocline, in the **profundal** zone, the physical and chemical features are more uniform, and abundance and biodiversity of animals considerably lower. The fauna is generally dominated by oligochaete worms, unionid mussels (Fig. 3.11), or insect larvae, such as chironomids (Fig. 3.20).

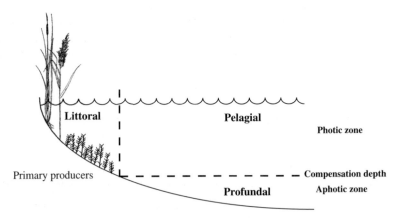

Fig. 2.21 Different zones of lakes and ponds, including the 'open water' (pelagic) zone and the nearshore littoral zone, where the primary production at the sediment surface is high. Water and sediment below the compensation depth (i.e. where the amount of oxygen produced by photosynthesis equals that consumed by respiration) is called the profundal zone.

Animal bioturbation may mix sediment particles in different ways. Oligochaete worms have their head down in the sediment and excrete faeces on the sediment surface, causing an upward movement of particles. On the other hand, many chironomids feed at the surface and excrete deeper down, thereby causing a net downward transport. Resting cells and eggs of many zoo- and phytoplankton species deposited in the sediment are redistributed by bioturbating animals. Bioturbation may thus transport such resting stages either deeper down in the sediment or closer to the surface and thereby change the conditions and make the resting cell hatch and enter the water column. An experiment showed that in the presence of the water louse (*Asellus aquaticus*), which moves around arbitrarily like a bulldozer in the sediment, the alga *Anabaena* (Fig. 3.6) hatched from the sediment in high abundance and rapidly formed a dense population in the water column (Ståhl–Delbanco and Hansson 2002). On the other hand, chironomids, which feed carefully at the sediment surface, had no effect on algal recruitment to the water column (Fig. 2.22). This study illustrates that different animals have different efffects on the hatching of resting cells, and, further, that bioturbation may be of importance in the initiation of algal blooms.

Bioturbation not only affects particles, but also the transport of nutrients through the sediment–water interface. With respect to both phosphorus and nitrogen, the activity of animals at the sediment surface generally increases the flux between sediment and water (Andersson *et al.* 1988). Because of high rates of decomposition by bacteria, sediments often have lower oxygen concentrations than the overlying water. The activity of bioturbating animals increases the oxygen penetration into the sediment. For example, the burrow of a chironomid larva is surrounded by oxygenated sediment, even if the sediment as a whole is depleted in oxygen. This oxygenated microlayer is a result of the larva transporting oxygenated water into the burrow by respiration and feeding movements. Such oxygen input from one single larvae may not be quantitatively important, but since

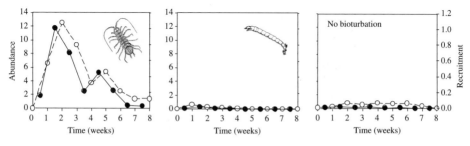

Fig. 2.22   Recruitment from the sediment (No. of algae 10^6 m^{-2} day^{-1}; filled symbols) and abundance (No. of algae x 10^3 l^{-1}; open symbols) in the water of the cyanobacteria *Anabaena* sp. in an experiment with *A. aquaticus* and *C. plumosus* as bioturbating animals, compared with a control treatment without sediment dwelling animals.

the abundance of chironomid larva often reaches several thousand individuals m^{-2}, they may have a significant influence on the oxygen concentration, and thereby on the chemistry, of the surface sediments.

Oxygen

Oxygen is, of course, a necessary requirement for all organisms with **aerobic respiration** ('breathing'), which include the majority of freshwater species. The seasonal and spatial changes in the availability of oxygen affect their life history strategies, distribution patterns, behaviour, and interactions with other organisms. The major inputs of oxygen in ponds and lakes are through diffusion from the atmosphere and from the release by plants during photosynthesis. Oxygen in freshwaters is consumed by organisms with aerobic respiration during the complex biochemical processes of catabolism where nutrient molecules (carbohydrates, fats, and proteins) are broken down, hydrogen is released, and then combined with oxygen. The energy contained in the chemical bonds of the nutrient molecules is transferred to energy-rich ATP molecules which are used by the organism for functioning and biosynthesis. The process of aerobic respiration can be simplified as:

$$C_6H_{12}O_6 + 6O_2 \xrightarrow{\text{enzymes}} 6CO_2 + 6H_2O + \text{energy (Respiration)}$$

and as can be seen, carbon dioxide and water are by-products in this process. Although the majority of organisms on earth, including yourself at this very moment, rely on aerobic respiration, some organisms can survive a complete lack of oxygen through alternative biochemical pathways (**anaerobic respiration**, see below).

Factors affecting the oxygen concentration

The amount of oxygen that water can hold in solution decreases with increasing temperature, which is, of course, disadvantageous for organisms because the metabolic rate, and thus the oxygen demand, increases with increasing temperature. The amount of oxygen in solution at any given time is sensitive to changes in the rates of physical processes, such as *mixing* and *wave action*, and biological processes, such as *respiration* and *photosynthesis*. For example, during periods of high photosynthetic activity, water becomes supersaturated with oxygen, whereas when decomposition processes predominate, oxygen levels are considerably reduced. Thus, the actual amount of oxygen dissolved in the water may deviate from the theoretical maximum, and therefore the amount of oxygen in water is often expressed as a percentage of the theoretical value for a given temperature (percentage saturation). Macrophytes and algae only produce oxygen in the presence of light which leads to that during night, or at depths where light levels are low, respiration reduces the oxygen concentration. The specific light level where the rate of photosynthesis (P) is similar to the respiration (R) rate; that is $P = R$, is called the light **compensation point** (Fig. 2.21). At higher light levels $P > R$. The physical properties of water affect the availability of oxygen for aquatic organisms, especially important being the slow diffusion rate and the low solubility of oxygen. The large temporal and spatial variations in oxygen content have important consequences for the organisms that depend on oxygen for their respiration and have, as we will see, led to many aquatic organisms having efficient adaptations to acquire it.

Estimating oxygen production and consumption

The rate of photosynthesis performed by algae and macrophyte plants affects the oxygen concentration in the water and is, as we have seen, itself affected by several factors, such as temperature and the availability of light and nutrients. The oxygen concentration can be measured either by an oxygen electrode, or by the somewhat old-fashioned, but highly reliable, *Winkler titration*. (For description of the method, see Wetzel and Likens 1991).

In the presence of light, photosynthesis (production of oxygen) and respiration (consumption of oxygen) both occur simultaneously and together determine the oxygen concentration of the water. The rates of photosynthesis and respiration can be determined by calculating the difference in oxygen concentration between a clear ('light') and a dark (e.g. wrapped in aluminium foil) bottle that are allowed to hang in the water. After determining the initial oxygen concentration, the bottles are left for, say, 1 h and then taken up and the oxygen concentration in each bottle is determined either with the Winkler method or with an oxygen electrode. The *net photosynthesis rate*, or the *primary production*, is the increase in oxygen concentration in the light bottle compared to the initial sample. Similarly, the *respiration rate* is the decrease in the dark bottle compared to the initial oxygen concentration. The total, or *gross photosynthesis rate*, is then the sum of the net photosynthesis and the respiration rates.

Spatial and temporal patterns in oxygen availability

Seasonal stratification

Seasonal changes in temperature and the resulting stratification patterns (Figs 2.5 and 2.19) are crucial for understanding the dynamics of dissolved oxygen in ponds and lakes. During spring, when the water starts to warm up, the differences in water density with depth even out, the resistance to mixing disappears, and the lake will circulate when the weather is windy. Oxygen-rich surface water is then distributed throughout the entire water column and saturation levels are at, or close to, 100% (Fig. 2.5). With increasing temperatures in summer, the lake will stratify into two stable layers of water (epilimnion and hypolimnion) that do not mix with each other. *Atmospheric oxygen* will be dissolved and distributed through diffusion and mixing within the epilimnetic layer, but there will be no transport of atmospheric oxygen down to the hypolimnion. The second major source of oxygen in lakes, *photosynthetic oxygen*, is completely dependent on light availability, which decreases with depth. Below the photic zone, light is insufficient to allow net-photosynthesis and oxygen production. The decomposition of organic material by bacteria in the hypolimnion or at the sediment–water interface, together with the respiration of benthic invertebrates, consumes significant amounts of oxygen and as there is no input, oxygen levels in the hypolimnion will decrease during the summer. With the increasing productivity of eutrophic lakes, the amount of organic material in the hypolimnion will be larger and the rate of oxygen depletion faster and, eventually, oxygen will be completely depleted in the hypolimnion. Thus, the hypolimnion will be unavailable as a habitat for most animals and plants during a large part of the summer.

In autumn, the amount of solar energy reaching the lake is reduced and water temperatures will decrease. Eventually, the lake water will overturn and oxygenated water circulates down to the deeper strata (Fig. 2.5). At the formation of an ice cover during winter, the exchange of oxygen with the atmosphere will be blocked. If the ice is transparent, there will be a considerable production of oxygen by photosynthesizing algae immediately under the ice, whereas in deeper layers oxygen-consuming decomposition processes will dominate. The amount of dissolved oxygen will thus decrease with increasing depth during the winter and be particularly low close to the bottom (Fig. 2.5). If the ice is covered by a thick layer of snow, photosynthesis and oxygen production will be almost completely suppressed because of the lack of light. If this continues for a long period the oxygen in the lake may be completely depleted, resulting in massive fish mortality. This is called '*winterkill*' and is especially common in shallow, productive ponds and lakes where decomposition of large quantities of dead organisms consumes a lot of oxygen.

Oxygen fluctuations in shallow waters

Organisms living in the littoral zone may also experience oxygen deficits during summer. In shallow areas with a high density of primary producers, such as submerged macrophytes and substrate-associated algae, dissolved oxygen levels change following a diel cycle. During daytime the photosynthetic activity is high in this well-lit habitat, resulting in high production of oxygen and saturation levels above 100%. During night-time, when the plants become 'animals', at least in the sense that they consume oxygen by respiration, dissolved oxygen in dense stands of submerged macrophytes may be severely reduced. In addition, in late summer when temperatures in shallow areas are high, decomposition processes may be so intense that plants start to die and oxygen can be completely depleted, resulting in catastrophic die-offs, especially of fish. This phenomenon is termed '*summerkill*'.

Morphological adaptations to obtain oxygen

Most freshwater organisms are dependent on oxygen for their existence, but as we have seen, oxygen levels and diffusion rates of oxygen in water are low and there are considerable seasonal and temporal variations in oxygen supply within lakes and ponds. To make up for this, aquatic organisms have evolved a number of adaptations that increase the efficiency of oxygen uptake. Depending on where the organisms acquire their oxygen, these adaptations can be divided into two categories: adaptations to use *oxygen dissolved in water* and adaptations to use *atmospheric oxygen*.

Oxygen dissolved in water

Most freshwater organisms are dependent on so-called **integumental respiration**, which means that oxygen is taken up directly across the body surface without any specialized respiratory morphological adaptations (Graham 1988, 1990). The most important feature favouring exchange of oxygen across the body surface is *small size*, resulting in a high surface area to volume ratio (SA : V). If the organisms are sufficiently small (zooplankton, phytoplankton) or thin (flatworms, leeches), diffusion rates across the body surface are efficient enough so that they can rely on integumental respiration and do not need any specialized respiratory structures (Fig. 2.23). However, the low diffusion rate of oxygen in water results in a diffusion boundary layer around the body surface which becomes depleted of oxygen. *Body movements* decrease the thickness of the oxygen-depleted zone and increase the availability of oxygen. The leech undulates its body to accelerate water movements, and the oligochaete, *Tubifex*, ventilates its body surface by extending its tail from its burrow and waving it. It makes a corkscrew motion and both the length (surface area) and the intensity of its motion increase with decreasing oxygen content. However, above a critical size, diffusion distance becomes too great and the ratio of tissue surface area to volume too small to permit delivery of enough oxygen across the body surface. Therefore, in many aquatic organisms special morphological structures have evolved to increase the efficiency of oxygen uptake.

Insects have a complicated tubular, **tracheal system** for transport of gases throughout the body (Fig. 2.23). In most aquatic insect larvae and nymphs,

Zooplankton   Insects   Fish

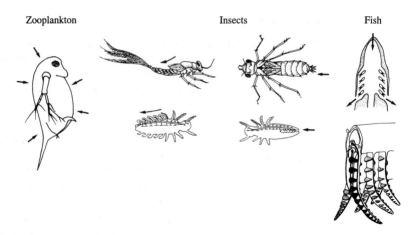

Fig. 2.23   Examples of freshwater organisms that use oxygen dissolved in the water. In small organisms, such as zooplankton, oxygen diffuses directly across the body surface. In insects, the tracheal system transports oxygen throughout the body. To increase surface area, the trachea may extend into external gills, such as in mayflies (left) or form a complicated network inside the rectum, as in dragonflies (right). Fish take up oxygen from water that is moved over their gills (above), which have filaments and lamellae to increase surface area (below).

the tracheal system is closed, and depends on diffusion across the seal. To increase surface area, the sealed tracheal systems may extend outside the body surface in the form of elaborate tracheal **gills**, such as the gill plates and gill tufts of mayflies (Ephemeroptera), or from a network of tracheae inside the rectum (dragonflies, Anisoptera) (Fig. 2.23).

Ventilatory movements increase the flow of oxygenated water across the respiratory organs. Free-living mayflies rapidly move their gills back and forth, whereas burrowing mayflies and chironomids and insects living in a case (e.g. caddisflies) create a water current through their burrow or case and over the gills by undulating the body. Anisopteran dragonflies pump water in and out of their richly tracheated rectal chamber at a rate of 25–100 movements per minute depending on oxygen availability. This **rectal pump** may also be used in jet-propelled swimming.

The surface area of fish gills is enlarged through their specialized morphology. The gill consists of a number of gill arches, each of which has numerous gill filaments, which, in turn, have secondary lamellae to increase the surface area even more (Fig. 2.23). The high density and viscosity of water results in fish having to spend considerable amounts of energy to move water across the gills. Highly active, fast-swimming species ventilate their gills by only opening the mouth so that water can enter and pass over the gills and leave through the opercular slits. Less-active fish have special respiratory movements, pumping water over the gill surfaces.

Oxygen from air

A number of freshwater animals are dependent on atmospheric oxygen for respiration. Many insects have special morphological structures that facilitate the uptake of air. *Respiratory siphons* ('snorkles') that can be opened at the surface are found in several insects, such as the water scorpion, many water bugs, mosquitoes, and soldier flies (Fig. 2.24) and large, tracheal trunks may serve as air stores and enable the insects to stay under water for long periods. Pulmonate snails visit the water surface to take air into the mantle cavity which is richly supplied with blood vessels and functions as a lung. Other insects *store air* beneath wing cases or trap an air bubble with hydrofuge (water-repellent) hairs that cover parts of the body surface. The hydrofuge hairs are bent over at the tip and enclose a layer of air that cannot be replaced by water. All air bubbles that are carried by organisms down into the water function as a physical gill if they are in contact with a respiratory surface of the animal. As the oxygen in the air bubble is consumed by the animal, it is replaced by oxygen from the surrounding water mass by diffusion, and thus the bubble supplies the animal with much more oxygen than was originally present. This process is called *plastron respiration*. The aquatic spider *Argyroneta aquatica* in addition to carrying a large air bubble trapped on the back of its hairy body, also has an air bell trapped in its submerged nest (Fig. 2.24). However, there are also costs associated with the use of atmospheric oxygen. For example, an animal swimming from deep water feeding areas to the surface to get air uses considerable amounts of energy and also exposes itself to a greater risk of predation.

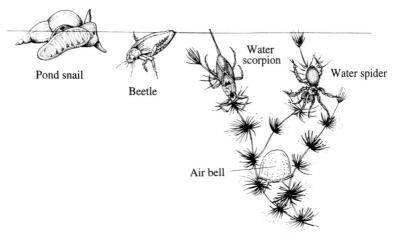

Fig. 2.24   Examples of organisms that use atmospheric oxygen for respiration. The pulmonate snail (left) takes air into its mantle cavity (the 'lung'), some beetles store air under their wing cases or form an air bubble with hydrofuge hairs. The water scorpion takes up air at the water surface with its respiratory siphon and the water spider (right) carries air on the back of its hairy body down to the submerged nest, the air bell.

Oxygen transport in plants

Oxygen supply is generally a less severe problem for photosynthesizing plants since they produce oxygen themselves during daytime. However, the photosynthesizing regions have to produce and transfer oxygen to the parts that are not green, such as roots and rhizomes. Emergent macrophytes have their photosynthesizing regions above the water surface, whereas their oxygen-demanding roots and rhizomes are buried in the sediment. To overcome this problem, stems and rhizomes are hollow, forming transport tunnels (**lacunae**) from the stomata in the leaves through the stem and down to the roots (Fig. 2.25). Some of the oxygen transport from emergent regions to the rhizomes and roots is driven by diffusion, that is, as the submerged parts respire, oxygen concentration becomes lower than in the atmosphere, leading to a flow of oxygen towards the rhizomes. However, the major part of the oxygen is supplied by '**internal winds**' created by the higher temperature ('thermal transpiration') and humidity ('hygrometric

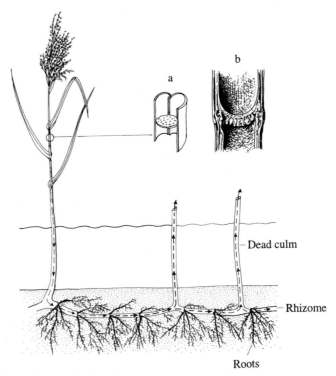

Fig. 2.25   A reed plant (*P. australis*) with rhizome and dead culms which together function as pipes for the 'internal wind' which supplies the rhizomes and roots with oxygen. The details show: (a) a schematic drawing of a cross-section of the hollow stem and the holes (the dots in the sieve-like plate) through which air can penetrate; (b) a similar cross-section showing how the sieve-like plate appears in reality.

pressure') inside compared to outside the plant, thus building up a gas pressure slightly higher than the atmosphere in young leaves. Older leaves and dead culms, which do not build up such pressure, instead function as 'vents' (Fig. 2.25). The 'internal wind' is highly efficient in providing oxygen to, and evacuating respiration products from, the root system when the sun is shining, but efficiency is low during night-time, leading to diurnal cycles in rhizome oxygen concentration. However, it is not light *per se* that drives the wind, but the increase in leaf temperature. This has been demonstrated experimentally by increasing and then decreasing the temperature of the plant in darkness, which created similar fluctuations in the gas transport as shifts between sunlight and darkness (Dacey 1981). Interestingly, the carbon dioxide (CO_2) evacuated from the roots is used by the plant in photosynthesis during daytime, thus recycling an often limited resource. 'Internal winds' have been demonstrated for the water lily (*Nuphar* sp.), where old leaves function as 'vents', and also for several emergent aquatic macrophytes, including common reed (*Phragmites australis*) and *Typha* sp. Plants that are able to create strong 'internal winds' can grow in deeper water and in more anoxic sediments than plants relying exclusively on diffusion.

Since some of the oxygen transported down to the rhizomes and roots diffuses to the substrate, the oxygen demand is higher in highly anoxic sediments ('mud') than in sandy substrates, due to a larger difference in oxygen concentration between the root system and the substrate. Moreover, the amount of oxygen reaching the root system decreases with increased distance between water surface and the sediment (water depth). This leads to the conclusion that the maximum water depth an emergent macrophyte can grow in is determined by the ability to support the root system with oxygen. Hence, emergent macrophytes, such as the common reed (*P. australis*) will grow in deeper water in sandy, often wind-exposed areas than in 'muddy', sheltered areas with low oxygen concentration (Weisner 1987).

Physiological and behavioural adaptations to low oxygen availability

The development of low oxygen concentrations in the hypolimnion or at the sediment surface is one factor resulting in a severe deterioration of living conditions for most freshwater organisms. Another is the formation of toxic substances, such as sulfides (e.g. H_2S). A sediment rich in sulfides is easy to detect even for the human nose—it smells like rotten eggs! In order to survive toxic substances, organisms have to overcome this environmental stress in some way or another. Some organisms simply leave the oxygen-depleted area and *migrate* to more oxygen-rich habitats, on a seasonal or even daily basis. Fish in winterkill lakes have been found to survive lethally low oxygen levels by migrating upwards to more oxygen-rich waters just beneath the ice or by leaving the lake by moving up into inlet or outlet streams that are oxygenated throughout winter (Magnuson *et al.* 1985).

Many aquatic organisms, especially invertebrates, have evolved specific adaptations that render them tolerant to lack of oxygen and allow them to thrive in oxygen-poor bottom waters and sediments for longer or shorter periods. In this way, they are to a large extent relieved of predation pressure by most fish, which are less tolerant to low oxygen levels. Several unrelated groups of freshwater organisms, such as crustaceans, chironomids, and snails, have evolved *haemoglobin* (Hb) as an adaptation to survive low oxygen. Hb is used by aquatic organisms in the same way as by humans (i.e. it facilitates transport, diffusion, and storage of oxygen). The concentration of Hb varies between closely related species depending on oxygen levels in their habitats, but there also exist differences within a species. Larvae of *Chironomus plumosus* that inhabited the oxygen-poor profundal had more Hb than larvae from the littoral zone where oxygen levels were high. In addition to chironomids, some pulmonate snails, such as planorbids, use Hb to increase their ability to withstand low oxygen levels. Some organisms can alternate between methods of obtaining oxygen depending on its environmental availability. Pulmonate snails acquire large amounts of oxygen through integumental respiration when dissolved oxygen levels are high, which enables them to stay submerged for long periods of time. As oxygen levels are reduced in the water, snails change to aerial breathing and the period of time between visits to the surface is a function of oxygen levels.

The majority of the protozoa, including amoebae, ciliates, and flagellates have *aerobic metabolism*. However, some have *anaerobic metabolism* and are therefore able to survive in habitats without oxygen. The metabolic pathways are not well understood, but a type of fermentation process is likely to be involved. The energy yield (ATP) in all fermentation processes is lower than in aerobic metabolism (i.e. growth is slower for organisms with anaerobic than those with aerobic metabolism). However, since few organisms are able to withstand anoxic conditions, competition is low in these often very nutrient-rich habitats, which may compensate for lower growth rates. Organisms with adaptations to withstand low oxygen concentrations are usually dominant and important in anoxic sediments and in specific man-made habitats, such as sewage treatment plants.

Habitat permanence

Not all freshwater habitats contain water all year. Temporary ponds are completely dry for longer or shorter periods. Some ponds only dry out for a few weeks in hot summers, whereas in others the dry period may last for several years. The timing of the dry period also differs among ponds. In *temporary vernal ponds*, the basins fill up in spring, dry out during summer,

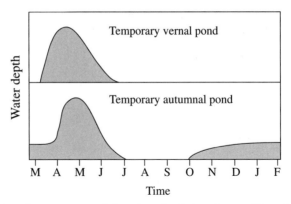

Fig. 2.26   Categorization of temporary ponds based on the timing of water-filling. Vernal ponds (above) get water-filled in the spring and then dry out in the summer, whereas autumnal ponds (below) re-fill in the autumn and remain water-filled throughout the winter. (After Wiggins *et al.* 1980.)

and do not get water-filled again until the next spring. *Temporary autumnal ponds*, on the other hand, dry out in the summer but refill in the autumn and remain water-filled all through the winter (Fig. 2.26). A typical temporary pond has no inlet or outlet and is dependent on precipitation and surface water run-off to be water-filled. Other ponds are merely isolated, that is, at high water levels in spring they are connected to nearby permanent streams or lakes and freshwater organisms can colonize by surface water connections. Most of the discussion below will focus on temporary vernal ponds.

Characteristics of temporary ponds

Temporary ponds may be considered a habitat with *harsh abiotic conditions*, a more or less unpredictable dry period obviously being the factor limiting the distribution of most freshwater organisms. Even when the pond is water-filled, environmental factors may set severe constraints on the organisms. For example, temporary ponds are typically *shallow* with a large surface area to volume ratio (SA : V) and this means that they are more susceptible to weather fluctuations than larger, deeper ponds and lakes. Generally, *temperature shows large diurnal fluctuations* and may even reach lethal levels as the water volume decreases. *Large fluctuations in water chemistry* are also common as a result of evaporation, rainfall, and dilution. Thus, organisms inhabiting temporary ponds should have special adaptations in order to be able to take advantage of this habitat. Nevertheless, temporary pools are used by an array of freshwater organisms, some of which are highly specialized for life in a temporary habitat, whereas others are merely generalists that occur in all kinds of freshwater habitats. Although life in temporary ponds may seem harsh and unpredictable, these habitats have many characteristics that make them highly favourable.

Most of the colonizers that first appear in a pond when it fills with water are detritivores or herbivores and they are provided with a rich food source. During dry periods, the pond basin is invaded by terrestrial vegetation that takes advantage of nutrients accumulated in the sediments and develops into rich stands. Decay of this terrestrial vegetation is enhanced by the aerobic conditions and by the fact that decomposing vegetation exposed to air has a higher protein content than detritus from water-filled habitats, that is, decaying vegetation provides a quantitatively as well as qualitatively rich food source for detritivores in spring (Bärlocher *et al.* 1978). Further, the combination of a high concentration of nutrients and excellent light availability in the shallow waters provide ideal conditions for a *rich algal growth*. Moreover, the predators that are important in regulating prey populations in permanent habitats are either absent (fish) or arrive late in the water cycle (invertebrate predators), leading to a *low predation pressure*. When predators arrive many organisms have already completed their life cycle and left the pond, or have reached a size where they are no longer vulnerable to predation. Thus, organisms that have adaptations to survive the dry periods have an advantageous habitat with rich food sources and low predation pressure, and can build-up high population densities. In the tropics, high population densities of invertebrates in temporary pools can be a serious problem, since several of these organisms are **vectors** for diseases, for example, snails that are hosts to *Schistosoma* trematodes causing Bilharzia in humans (see Chapter 4, *Biotics*), and mosquitoes that can transmit malaria and other severe tropical diseases.

Adaptations

Freshwater organisms that inhabit temporary ponds have a variety of adaptations to cope with this environment (Wiggins *et al.* 1980). Some are generalists that occur in a variety of freshwater habitats, but have the necessary adaptations for life in temporary waters, whereas others are highly specialized and only occur in temporary waters. The organisms could be broadly divided into two separate categories: in the first we have organisms that stay in the pond basin throughout their life cycle, whereas organisms that leave the pond at some stage are in the second category. Naturally, organisms in these different categories use somewhat different adaptations to enable them to inhabit temporary habitats.

Adaptations among permanent residents

Animals that are permanent residents in temporary ponds, independent of water level, typically find protection by *burrowing* into the bottom sediment, and they also have different structures, such as *resting eggs* and *mucous* coats, that make them resistant to desiccation. When the pond fills with water, these organisms can become active within a very short time. For example, increasing water temperatures as the pond recedes in spring cause some flatworms to fragment into small pieces, each of which encapsulates in a hard mucous cyst. During the dry period a young worm develops within the cyst and as soon as the pond refills it hatches and starts to grow. Similarly, oligochaetes and leeches burrow into the moist sediment and secrete a protective mucous layer, whereas pulmonate snails can breathe atmospheric air and protect themselves against desiccation by sealing the shell aperture with a mucoid epiphragm. Fairy shrimps (Anostraca) only occur in temporary waters, but they are widely distributed around the world. They survive the dry periods as resting eggs that are resistant to drought and can remain viable for many years. A complex interaction of abiotic variables, including high temperature, high oxygen level, and moisture, synchronizes the development of eggs so that they hatch in spring. Fairy shrimps are filter-feeders and utilize the abundant source of organic particles to grow rapidly and complete their life cycle before predatory insects are at their maximum population density. Some crayfish species can inhabit temporary ponds but only if they can dig burrows during the dry season that are deep enough to reach the ground water-table. Other organisms commonly take advantage of the moist environment in crayfish burrows and survive the dry season there.

Adaptations among temporary residents

Although many organisms have developed adaptations to cope with the dry phase within the pond, others *leave the pond* basin during some stage of their life. Many of these organisms have complex life cycles, which include an abrupt ontogenetic change in morphology, physiology, and behaviour, usually associated with a change in habitat (Wilbur 1980), that is, they **metamorphose**. This change in morphology may be necessary to enable utilization of the different ecological niches exploited by the two stages of a metamorphosing animal (Werner 1988). The different niches of the life stages naturally result in different selection pressures on each stage, and genetic correlations between morphological structures during a non-complex life cycle constrain the evolution of adaptations to the different niches. Thus, evolution of complex life cycles and metamorphosis may be a way of promoting independent evolution for the different stages to their specific niche, so the genetic covariance of morphological structures is disrupted (Ebenman and Persson 1988). Wilbur (1984) also suggested that metamorphosing organisms in temporary habitats are adapted to exploit a rich, transient resource that provides an opportunity for rapid growth, whereas the adult stage is adapted for dispersal and reproduction.

Many insect taxa use the rich resources of temporary ponds for rapid growth through their juvenile stages. However, there is considerable variation among species in their life history strategies and especially on how they survive the dry period. Some species, overwintering spring recruits, spend winter within the dry pond basin as eggs or adults. In spring they are capable of dispersal as adult insects. Oviposition is dependent on water and, thus, they must reproduce before the pond dries up. Overwintering summer recruits have their dispersal period in the summer and arrive at the pond after the water has disappeared (i.e. egg-laying is independent of the presence of water). Some caddisflies, for example, deposit their eggs in a gelatinous matrix in protected places in the dried-out pond during early autumn. The larvae develop and hatch from the eggs within the matrix after a few weeks, but stay there until water returns. Non-wintering spring migrants arrive at the water-filled pond in spring where they reproduce and oviposit in the water-filled basin. The juveniles metamorphose before the onset of the dry period and spend their adult life either as a terrestrial organism (e.g. some amphibians) or disperse to a permanent pond, where they may even complete another generation (e.g. some mayflies, Ephemeroptera). Hemipterans, such as water-striders and water-boatmen, may overwinter as adults in permanent ponds. In spring they colonize temporary habitats through flight, and oviposit. The juveniles grow rapidly in the pond, metamorphose before it dries up, and disperse as adults to a permanent waterbody. Thus, there is a high selection pressure for juveniles to metamorphose before the pond dries up. Amphibians are known to show a plasticity in their growth rate and size at metamorphosis. The timing of amphibian metamorphosis is a trade-off among different selection pressures. Larger size at metamorphosis may result in higher fitness, but staying in the pond to grow large also increases the risk of death through desiccation. It has been shown that amphibians can make a compromise between these demands and metamorphose at a smaller size in years when the pond dries up earlier.

Dispersal modes

Species colonizing temporary waters as they fill with water should have efficient means of dispersal. Many organisms are *active colonizers*, such as insects that fly between habitats or crayfish, leeches, and amphibians that crawl between ponds or colonize from a terrestrial habitat. Others, *passive colonizers*, have to depend on transportation by wind if they are very light (e.g. resting eggs) or with other active migrants. Most knowledge on means of dispersal in aquatic organisms comes from anecdotal evidence (e.g. observations on different organisms 'hitch-hiking' with waterfowl or water beetles). Further, **asexual reproduction, parthenogenesis**, or **hermaphroditism** should increase the probability of successful dispersal, because a single individual is then sufficient to start a new population.

We have now set the abiotic frame for organisms in freshwaters and also seen some examples of adaptations to cope with abiotic constraints. The examples have included several organisms, of which some may be well known, others not. Therefore, before we take the step into the intriguing world of interactions between organisms, we provide a more thorough presentation of freshwater organisms in the next chapter, *The organisms: the actors within the abiotic frame.*

Practical experiments and observations

Estimates of light penetration and algal biomass

Secchi depth

Background   As mentioned in the text, the Secchi depth is often used as an estimate for light penetration through water, which is mainly affected by particles in the water.

Performance   The principle is very simple: lower a white (or black and white) circular plate (Secchi disc) into the water until it disappears and then lift it until you see it faintly. Continue to lower and lift until you find the depth where the disc is just visible. The distance between the plate and the water surface is the Secchi depth (named after the Italian priest and astronomer P. A. Secchi who developed the method). Ensure that the reading is taken in a place without sun reflection from the water surface, for example on the shady side of the boat, as sun reflexions will interfere with the reading.

Chlorophyll concentration

Background   In the same way as Secchi depth is a rough estimate of light penetration, the concentration of chlorophyll *a* is an estimate of algal biomass. Generally, chlorophyll *a* constitutes between 2% and 5% of the dry weight of an algal cell.

Performance   Filter about a litre of lake water through a Whatman GF/C or corresponding filter. Write down the volume filtered, you will need it later. Put the filter in a test tube with 7 ml of solvent (e.g. ethanol, 96%) and seal the test tube with a stopper. Store it in the dark at room temperature overnight. Centrifuge the sample at 4000 cycles per minute for about 10 min. Measure the chlorophyll on a spectrophotometer at 665 nm, which is the absorption maximum of chlorophyll *a*, and at 750 nm, where chlorophyll has very low absorption, but other particles, which we are not interested in here, have high absorption. Calculate the chlorophyll concentration ($\mu g\,L^{-1}$) according to the following formula:

$$\text{Chlorophyll conc. } (\mu g\,l^{-1}) = ((\text{Abs}_{665} - \text{Abs}_{750}) \times e \times 10\,000)/(83.4 \times V \times l)$$

where *e* is the volume of the extract (in this case 7 ml), *l* is the length of the cuvette in the spectrophotometer (expressed in mm), *V* is the filtered volume (expressed in l), and 83.4 is the the absorption coefficient for ethanol (expressed in $l\,g^{-1}\,cm^{-1}$). If you use methanol as solvent the absorption coefficient is 77.0.

To discuss Measure Secchi depth and chlorophyll concentration in some lakes or ponds in your area. You may also gather data from the literature. Is there a relation between Secchi depth and chlorophyll values? If there is, why? Now, include lakes with high humic content into your dataset. What is the relation between chlorophyll concentration and Secchi depth in these lakes? Why?

Bioturbation

Background This experiment demonstrates how different benthic animals, such as oligochaetae worms and chironomid larvae, cause sediment material to mix in different ways.

Performance Cover the bottom of two small aquaria, or jars, with white sand. Then carefully cover the sand in both aquaria with about 1 cm of sediment (black), and, again very carefully, pour water into the aquaria without disturbing the two layers. Bring in lake sediment and pick out chironomids and oligochaete worms. Release the chironomids into one of the aquaria and the oligochaetes into the other. Wait a few days. What has happened in the two layers you so carefully put into the aquaria? Are there any differences between the aquaria? Why?

If you have access to benthic feeding fish, such as bream (*Abramis brama*) or crucian carp (*Carassius carassius*), you may use the same experimental set-up. What happens in the sediment surface? How is the water affected by the fish?

To discuss How is the oxygen concentration at 2 cm sediment depth affected by chironomids and oligochaetes, respectively?

Temperature gradients

Background The temperature in a pond or lake obviously changes over the seasons, but there are also some changes occurring from day to night. Further, different habitats in a lake differ in water temperature. These temporal and spatial changes obviously affect the organisms living in the waterbody.

Performance Go to a lake or a pond on a sunny spring day and measure the temperature along a depth gradient from the shallow, nearshore areas with lush macrophyte stands to the deep waters of the pelagic zone. Measure the temperature at 50 cm depth intervals at each sampling point along the transect. Do the measurements in the morning, midday, afternoon, evening, and night.

To discuss   Is there any difference in temperature along the depth gradient? Is the lake/pond stratified? Is the change in temperature different between different habitats? What are the consequences of temperature differences for the organisms inhabiting the waterbody? Speculate about where a female fish should deposit eggs in the spring in order for rapid development.

Exposure

Background   Wind exposure differs between areas in a lake and may have profound effects on the distribution and abundance of different organisms, plants as well as animals.

Performance   Select areas in your study lake that are either sheltered or exposed to wind. Investigate substrate characteristics and the occurrence of plants and animals along transects perpendicular to the shore in the different areas.

To discuss   How does wind exposure affect the distribution of littoral zone organisms? What adaptation do organisms have that are dominant in sheltered versus exposed areas?

Temporary ponds

Background   Temporary ponds are completely dry during parts of the year and this naturally has a dramatic impact on the living conditions for freshwater organisms inhabiting such ponds.

Performance   Select a nearby, temporary pond and monitor the changes in benthic macroinvertebrates over the year. Be sure to sample the pond (sweep-net) in the spring before the drying period and then at regular intervals when the dried pond gets water-filled again. Take sediment cores during the dry period, add water, and see what happens.

To discuss   Which animals are present in the pond during its water-filled stage? Are there significant differences compared to a permanent pond? Which animals colonize first after the dry period? Have they survived within the dry pond or do they have a terrestrial life stage? Which functional groups dominate during the water-filled stage (grazers, herbivores, predators)?

3 The organisms: the actors within the abiotic frame

Introduction

Systematics is a constantly evolving discipline, where units (species, genera, and other taxonomic groups) are revised and renamed at irregular intervals. Because most taxonomic work was performed several hundred years ago, an organism's taxonomic position is based mainly on characters visible to the human eye or with limited microscopic magnification. During recent years, new sophisticated techniques, including molecular biology and electron microscopy, have revealed new patterns of kinship that are based not so much on visible characters but, for example, on similarity in DNA sequences. The methodological and conceptual revolution has changed systematics from a scientific branch dominated by field naturalists, experts in their specific taxa, to a dominance by urban laboratory biologists using 'high-tech' molecular methodology.

Systematics has always been an important component of ecology, since it is of crucial importance to know what sort of organism you are working with, and there is no reason to believe that the need for taxonomic knowledge will decrease in the future. In this chapter, we will, however, base the relationships between organisms on their function in the ecosystem, thereby focusing less on pure taxonomic links, and more on size, shape, and relationships within aquatic food webs. Consequently, we use a simplified food web as the foundation of this overview. Some organisms will be covered in detail, whereas others are only briefly described. This partly mirrors their importance and commonness in lakes and ponds, and partly reflects the fact that the organisms are used as examples in other chapters of the book. Hence, this chapter is an overview where you can see what an organism looks like and get introduced to the intriguing variability and diversity among organisms in freshwater habitats. We will start with showing the size distributions of the major groups of organisms (Fig. 3.1). The approximate size ranges are shown with vertical bars.

The smallest organisms are the viruses, followed by bacteria which range in size from about 0.001 to 0.007 mm. The majority of aquatic organisms are in the size range 0.01–1 mm, including protozoans, rotifers, most algae, and many macrozooplankton. Organisms easily visible with the naked eye (i.e. larger than about 1 mm), include macrozooplankton, macroinvertebrates, many periphytic algae, as well as macrophytes, and fish (Fig. 3.1).

As already indicated, the classification of organisms is often old-fashioned and may sometimes have problems in acommodating new findings.

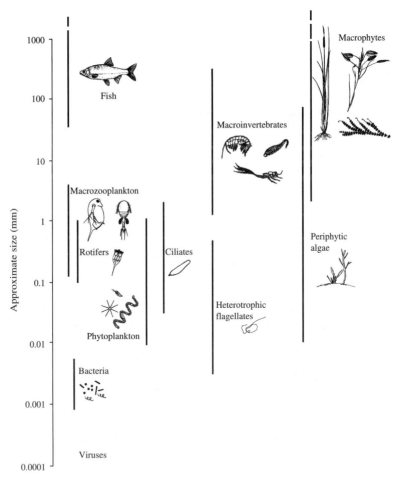

Fig. 3.1 The size spectrum of organisms in freshwater ecosystems, showing approximate ranges in size (vertical lines) of groups of organisms from viruses (less than 0.001 mm) to fish and macrophytes that may be larger than 1000 mm. Note that the scale is logarithmic.

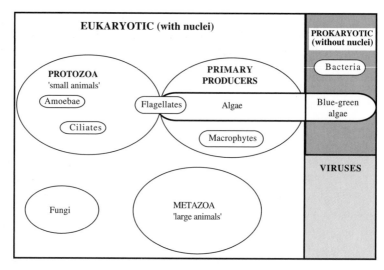

Fig. 3.2   Overview of commonly used classifications. Since the definitions have been created for practical reasons, there are several overlaps and many organisms are difficult to classify. For example, the border between prokaryotic (no nuclei) and eukaryotic (with nuclei) cuts right through the algae, leaving cyanobacteria with the prokaryotic, whereas the rest of the algae remains with the eukaryotic organisms. Similarly, as the difference between protozoa and algae is heterotrophic or autotrophic (photosynthesis) nutrition, the heterotrophic and photosynthesizing flagellates are categorized as protozoa and primary producers, respectively.

Despite this, taxonomic classifications are still widely used and a brief knowledge of the terms is of great help when reading scientific books and papers. In Fig. 3.2, the overall classification of organisms is shown in a condensed form. Here, organisms are divided into those that have nuclei (eukaryotic organisms) and those that have not (prokaryotic organisms and viruses). Eukaryotes include protozoa ('small animals'), plants, metazoa ('large animals'), and fungi (Fig. 3.2). Prokaryotic organisms are bacteria and cyanobacteria. Viruses were discovered after the classification system was developed and constitute a group of their own.

In addition to a taxonomic categorization, freshwater animals are commonly divided into different groups based on their ecology (e.g. filter-feeders, predators, planktonic, benthic), or their size. For example, benthic invertebrates are often divided into macroinvertebrates, meiofauna, and microbenthos. The basis for this classification is simply the mesh size of the sieves used when processing the samples. Macroinvertebrates are defined as the invertebrates retained by sieves with a mesh size larger than 0.2 mm, meiofauna is usually sieved with 0.1 mm sieves, whereas the microbenthos are the organisms that pass through sieves larger than 0.04 mm.

Below, we give a condensed presentation of the organisms in lakes and ponds. It is not intended to cover all aspects of aquatic life, but we try to provide some general information on morphology, life cycle, and feeding for all groups. You may use this chapter either to scan, and be fascinated by, the enormous variation of aquatic life, or use it as a lexicon when you need information on any specific organism or group.

Viruses

Viruses are small (usually between 20 and 200 nm) obligate intracellular parasites. They consist of genetic material only (DNA or RNA) covered by a protein coat and can be viewed as something in between an organism and a molecule, since they lack some of the features common to 'normal' living organisms. For example, they cannot reproduce without help from a host cell. Viruses (the word means 'poisonous substance') were discovered much later than other organisms, and they do not fit very well into our classification systems and therefore form their own group (Fig. 3.2). The knowledge of their role in aquatic ecosystems is still negligible, although it is clear that viruses play an important role in the population dynamics of freshwater communities by infecting organisms ranging in size from bacteria to fish.

Life cycle: Viruses enter their host cell by using its transporter proteins through the cell membrane. When inside, the virus may use several types of reproduction. One of the most well known is *lytic infection* where the virus injects its nucleic acid (DNA, RNA) into a host cell, programming it to produce new viruses. The viruses eventually become so numerous that the cell bursts (**lysis**) and releases the viruses which then find new host cells to infect. Another way for viruses to reproduce is to infect its host by *chronic infection*; that is, the host cell is infected in the same way as described above, but it does not die. Instead, it continuously releases newborn viruses over several generations.

The abundance of viruses in natural waters is generally about 10^7 ml^{-1}; that is, they are about ten times more abundant than bacteria. However, abundances vary widely and, as for most other organisms, viruses are more abundant in eutrophic than in less productive waters, and more common during summer than during winter. Fluctuations in virus abundances are very rapid and doublings in numbers have been recorded in less than 20 min (Wommack and Colwell 2000). High abundances of viruses are generally directly connected to high abundances of bacteria or to algal blooms. Despite difficulties in quantifying how efficient viruses are in infecting their host organism, rough estimates indicate that between 10% and 20% of the bacteria and 3–5% of the phytoplankton are killed per day by virus infections (Suttle 1994). Carbon and other nutrients released from lysed cells rapidly become available and can be reused by bacteria and algae, thereby forming a subset of the 'microbial loop' (Fig. 5.14; Fuhrman 1999;

Thingstad 2000). In addition to having an impact on the quantity of organisms, virus may also affect the species composition and biodiversity of, for example, bacteria and algae in lakes and ponds. Hence, any bacterial strain or algal species that become abundant will have a higher encounter rate with a, often relatively host-specific, virus. Once the host achieves a critical density, an epidemic of viral infection will start, selectively restricting the dominance of this fast-growing host, thereby reducing its competitive impact on other host strains or species and increasing biodiversity.

Prokaryotic organisms

True organisms are divided into prokaryotic and eukaryotic organisms, which lack and have, respectively, nuclei and other cell organelles (Fig. 3.2). Prokaryotic organisms include bacteria and cyanobacteria, whereas all other organisms (except viruses), are classified as eukaryotic.

Bacteria

Bacteria vary greatly in shape, including coccoids, rods, spirals, and chains. They generally range in size between 0.2 and 5 μm, which makes them appear only as 'dots' in a light microscope. Despite their small size, bacteria are extremely important in lake metabolism by being involved in mineralization processes and in the chemical transformation of elements between reduced and oxidized forms (e.g. with respect to nitrogen and sulphur). At optimal temperatures and resource availabilities, they may have generation times as short as 20 min. Usually, only a minor portion of the total number of bacteria in lake water is metabolically active; the majority are waiting for suitable conditions.

Bacteria may be viewed as the most abundant food organisms for higher organisms, occurring in very high numbers (usually 10^6 ml^{-1} water). They occupy an important position in the aquatic food web since they are major actors in the decomposition of dead material, and thereby in the recycling of nutrients and carbon. Despite their importance, little is known about their population dynamics and their response to grazing. The taxonomy of bacteria is still restricted to separation into different morphologies, including small single cells (<1 μm), larger cells (often rod-shaped), and filamentous forms. Bacteria are used as food for various grazers, such as rotifers, crustacea, and protozoa, such as ciliates and heterotrophic flagellates. Heterotrophic flagellates are similar in size and shape to some flagellated algae (e.g. *Cryptomonas*), but lack chlorophyll and therefore cannot fix their own carbon through photosynthesis. This forces them to consume other organisms. Ciliates and heterotrophic flagellates mainly graze on the medium-sized (rod-shaped) bacterial fraction, which, together with the filamentous forms, also are preferred by cladocerans.

Cyanobacteria lack nuclei and formally belong to the prokaryotic organisms and should consequently have been presented here. However, since they are generally treated as 'algae', they are included in the section on algae (see below).

Eukaryotic organisms

Fungi

Fungi are simple eukaryotic organisms lacking chlorophyll. Some use organic substances as substrates and are called **saprophytic**, whereas others are parasitic. Some are parasitic on higher organisms, such as *Saprolegnia*, which may completely cover an individual fish, whereas others attack algae and may be an important factor causing declines of algal blooms. Fungi are, like bacteria, involved in mineralization, especially on surface areas and in the sediment.

Protozoa

Protozoa are unicellular eukaryotes distinguished from algae by obtaining energy and nutrients by **heterotrophy** instead of by photosynthesis (i.e. by consuming complex organic molecules or particles, such as bacteria). Protozoa means 'first animals', a name they have because they employ the same feeding strategy as animals. Protozoa have very high reproductive potential with generation times of only a few hours in optimal conditions, allowing them to respond rapidly to resource fluctuations. Owing to their small size, they may also exploit very small patches of resources. Protozoa may be *amoeboid, flagellated,* or *ciliated.* Heterotrophic flagellates and ciliates have motility organelles (flagella and cilia; Fig. 3.3), allowing them to move at speeds of up to 0.1 and 1 mm s^{-1}, respectively.

Amoebae

Amoebae lack cell walls and move using temporary extensions (pseudopodia, Fig. 3.3, i.e. they turn their entire body into organs of locomotion). Food particles either attach to the sticky surface and are absorbed, or are enclosed by the pseudopodia.

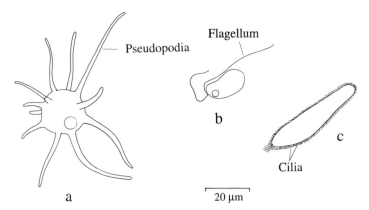

Fig. 3.3   Schematic drawing of: (a) a planktonic form of amoeba with pseudopodia; (b) a hete-rotrophic flagellate with two flagella; and (c) a ciliate, showing many cilia.

Ciliates

Ciliates is a successful group of organisms common in most freshwaters (Fig. 3.3). Their length is generally 20–200 μm, but some are more than 1000 μm long. For locomotion and feeding, ciliates use cilia, which are short 'hairs' that usually occur in high numbers on their cell surface. **Feeding:** ciliates mainly feed on bacteria, but also on algae, detritus, and other protozoans. Some are *mixotrophic* (i.e. supplement their feeding by photosynthesis). **Life cycle:** ciliates use a sexual reproduction called conjugation in which two cells fuse together in the mouth region allowing exchange of genetic information resulting in new individuals with genotypes differing from their parents. Many ciliates are able to grow at low oxygen concentrations, making them successful organisms in polluted water. One well-known example of this is the genus *Paramecium* (Fig. 3.3).

Flagellates

Flagellates are protozoans that use hair-like structures (flagella) for loco-motion. In contrast to the cilia of ciliates, there are fewer (1–8) flagella and they are usually longer than the body. Flagellates may be heterotrophic, which means that they feed on other organisms and are therefore classified as animals. Other flagellates have chloroplasts, which means that they are partly autotrophic (i.e. they use photosynthesis for nutrition). To add to the confusion, some of the autotrophic flagellates are able to ingest particles in addition to retrieving the major part of their carbon from photosynthesis; they are both animals and plants! Hence, although the science of systematics draws a rigid border between plants and animals based on the occurrence of chloroplasts, nature itself exposes a range of organisms from pure animals to pure plants. **Feeding:** heterotrophic flagellates feed mainly on bacteria and are quantitatively important in reducing bacterial abundance.

Primary producers

Primary producers rely on photosynthesis for their nutrition. The main feature of primary producers is the green colour caused by the pigment chlorophyll, which is a major factor in gathering energy from sunlight. Aquatic plants may be divided into macrophytes, phytoplankton (free-living), and periphytic (substrate-associated) algae. All primary producers have basically the same resource requirements, including nutrients (e.g. phosphorus and nitrogen), carbon dioxide, and light, although they have different solutions and adaptations to optimize resource intake.

Macrophytes

The main advantages for aquatic macrophytes compared to their terrestrial relatives are first that they do not depend on water supply, and second that water is viscous enough to stabilize stems and leaves. This reduces the need for support tissue, thereby releasing energy that can be used for faster growth.

Emergent macrophytes photosynthesize above the water surface and have aerial reproductive structures. Some emergent macrophytes form dense monocultures in shallow water, including common reed (*Phragmites australis*, Fig. 3.4). **Life cycle:** reed stands are able to expand rapidly due to below-ground rhizomes from which new shoots grow. Sexual reproduction and concomitant dispersal of seeds are generally less important since the seed and the new plant have specific requirements in order to germinate and grow, including optimal temperature and humidity regimes. Some emergent macrophyte species have seeds that can survive for many years in sediments, forming a 'seed bank' which may constitute a security against catastrophic events. A situation when seed dispersal is of importance for the expansion of emergent macrophytes is when the water level of a lake or pond is reduced (e.g. during drainage to increase agricultural areas). In such cases, the whole area with humid sediment left behind by the retreating lake may be colonized by germinating reed plants.

Submersed macrophytes grow primarily under water and are attached to the substrate (Fig. 3.4). *Floating-leaved macrophytes*, including *Potamogeton* and *Nuphar* (water lily), are attached to the substrate, but have leaves that float on the water surface. *Free-floating macrophytes* are not attached to a substrate but float freely at the water surface, including *Lemna* (duckweed).

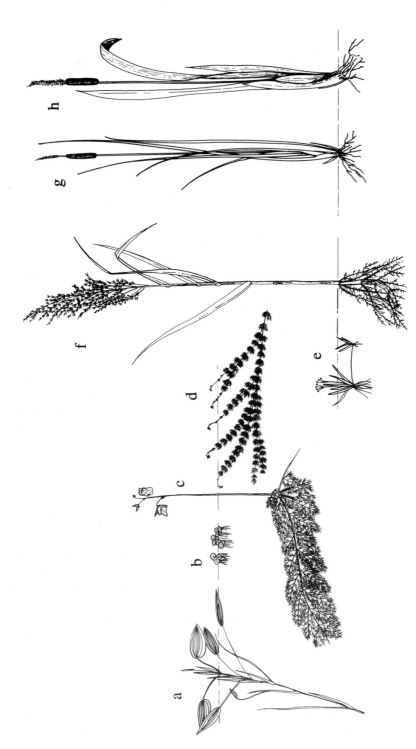

Fig. 3.4 Different growth forms of aquatic macrophytes: (a) is an example of a floating-leafed macrophyte (*Potamogeton natans*); (b) is a free-floating macrophyte (*Lemna* sp.); (c–e) are submersed macrophytes (*Utricularia* sp., *Myriophyllum* sp., and *Littorella* sp., respectively; (f) *P. australis*; (g) *Typha angustifolia*; and (h) *Typha latifolia* are representatives of emergent macrophytes.

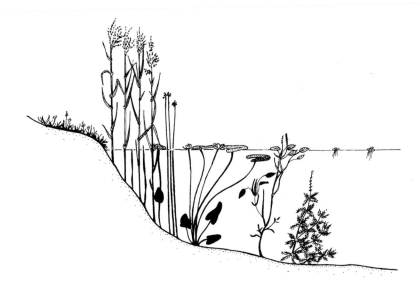

Fig. 3.5   A cross-section of a lake showing a typical zonation of macrophytes. Emergent macrophytes, such as *Phragmites* and *Typha*, grow closest to the shore. Further out, floating-leafed macrophytes become dominant, here represented by water lily (*Nymphaea*) and *Potamogeton*. Thereafter, submersed, and free-floating forms take over.

The different life forms of macrophytes generally occur at different depths, often constituting zones from the shore and out to deeper water. This zonation can easily be seen in most lakes and generally begins with emergent macrophytes close to the shore, followed by floating-leaved, and submersed macrophytes (Fig. 3.5).

Algae

Algae are probably the most well-studied group of all aquatic organisms and comprise a diverse range in size and shape, from single-celled species a few micrometres (μm) long to colonial or filamentous forms that can easily be seen by the naked eye. Algae range from simple, circular shapes, such as many centric diatoms, to highly specialized spined or flagellated varieties (Fig. 3.6). The variability and diversity make algae common and successful life forms in all lakes and ponds, since there are always some species that 'fit' in a specific constellation of environmental features. Algae occur both as free-living phytoplankton and as periphytic algae associated with a substrate.

Phytoplankton

'Phyto' means 'plant' and 'plankton' has its roots in the Greek word for 'wandering'. Hence, phytoplankton means 'wandering plant', which well describes both their nomadic way of living within a lake or pond and their

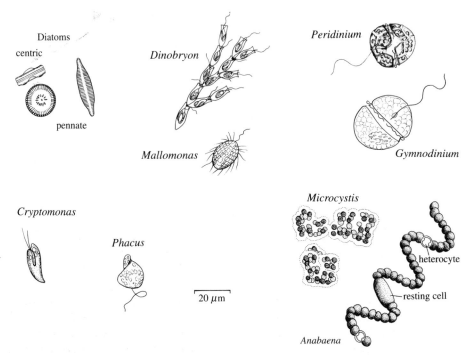

Fig. 3.6   An indication of the variations in shape and size among freshwater algae, showing pennate and centric diatoms (class:Bacillariophyceae), *Dinobryon* and *Mallomonas* (class: Chrysophyceae, 'golden algae'), *Cryptomonas* (class: Cryptophyceae), *Microcystis*, and *Anabaena* (class: Cyanophyceae, cyanobacteria). Note that *Anabaena* has a heterocyte where nitrogen fixation takes place. *Phacus* (class: Euglenophyceae), *Gymnodinium* and *Peridinium* (class: Dinophyceae). Note that *Peridinium* is covered with hard plates, whereas *Gymnodinium* is 'naked'. The scale bar represents 20 μm and applies to all algae in the figure.

wide global distribution. The algal cell is surrounded by a cell wall which is often made of cellulose and other polysaccharides, but may also contain proteins, lipids, and silica. The chemical composition of the cell wall is species-specific and has been used for taxonomic classification of algal species.

As we shall see in the following chapters, some phytoplankton have specific adaptations to increase the handling time for their enemies—the herbivores—enabling them to withstand high grazing pressure. Such adaptations include large size (e.g. *Ceratium, Aphanizomenon*), spines (*Staurastrum, Ceratium*), and mucous sheets thick enough to allow viable passage through the gut of a herbivore (e.g. *Sphaerocystis*, Fig. 3.7). Since specific adaptations generally have their costs, it may sometimes be more beneficial to be small and spend most energy on growth and reproduction,

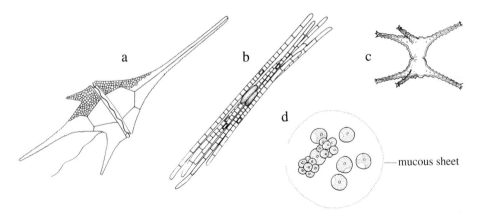

mucous sheet

Fig. 3.7   Adaptations among algae to make them resistant to grazing zooplankton:(a) large size and thick cell wall (*Ceratium* sp.); (b) 'spaghetti shape' and bundles (*Aphanizomenon* sp.); (c) spines in three dimensions (*Staurastrum* sp.); and (d) a protective mucus sheet allowing the cells to pass through the gut of the zooplankton unaffected (*Sphaerocystis* sp.).

instead of on specialized defence adaptations. Examples of such strategies are small algae without any protection, such as *Cryptomonas* (Fig. 3.6) and *Chlamydomonas*. It is important to note that different adaptations are successful at different times and in different environments, so that different algal species become dominant at different times and places. Hence, the species composition of the phytoplankton community is not arbitrary, but allows the alert observer to discover a lot regarding the characteristics of the lake or pond investigated, including nutrient status, climate, acidity, and the composition of the grazer community. Nutrient status (productivity) and pH are particularly important in determining dominance within the algal community, and the major phytoplankton groups may be characterized along gradients of pH and productivity (Fig. 3.8). As a rule of thumb, the phytoplankton community in low productivity (oligotrophic) lakes is often dominated by desmids if the pH is low (acidic), and by diatoms, such as *Cyclotella* sp. and *Tabellaria* sp., if pH is high (alkaline). In mesotrophic conditions, and at pH around neutral (7), chrysophytes and dinoflagellates are often abundant, whereas diatoms, such as *Asterionella* sp. and *Stephanodiscus* sp., may become abundant at high pH (Fig. 3.8). In high productivity (eutrophic lakes) and high pH, cyanobacteria are likely to dominate. **Feeding:** algae absorb nutrients directly from the water through the cell wall. **Life cycle:** most algae rely primarily on asexual cell division for their reproduction, although simple forms of sexual reproduction occur. Asexual reproduction may occur by simple cell division or by autospore formation where a new cell/colony is formed within an old cell, such as in *Scenedesmus* (Fig. 3.9). Below we briefly describe the characteristics of some important groups of algae.

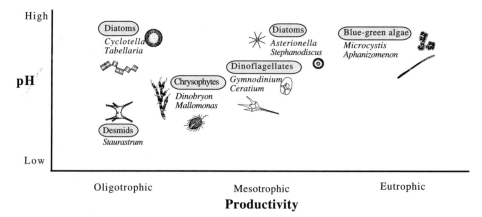

Fig. 3.8   The relative dominance of different algal groups in relation to productivity and pH. Examples of common genera likely to dominate are also shown. Knowledge about the composition and dominance pattern within the phytoplankton community may thus give an indication of the abiotic features of the specific lake or pond.

Cyanobacteria ('blue-green algae', 'cyanoprokaryotes')   Cyanobacteria are a primitive group lacking a nucleus and other cell structures and are therefore classified as prokaryotic (Fig. 3.2). Despite this, we present them here as they are always included in algal sampling. There has been, and still is, a debate regarding whether this group of organisms should be called 'cyanobacteria' or 'blue-green algae', or something else. Recently, the name 'cyanoprokaryotes' has been suggested. Throughout this book, however, we will use the name 'cyanobacteria'.

In addition to the lack of nuclei, cyanobacteria also lack chloroplasts, the structure that contains the pigments in other algae. Instead, the photosynthetic pigments are concentrated in an open lamellar system, not separating the pigments from the remainder of the cytoplasm. Cyanobacteria have the pigment, phycobilin, which gives them their characteristic bluish tinge. The majority of the cyanobacteria are filamentous including many well-known nuisance algae dominant in eutrophic lakes, such as *Aphanizomenon* (Fig. 3.7). Many filamentous cyanobacteria are able to form heterocytes, which are cells with thick walls where oxygen concentration is low, creating optimal conditions for nitrogen fixation (Fig. 3.6). Other cyanobacteria are unicellular or form colonies, such as *Microcystis* (Fig. 3.6).

Green algae   Green algae are a morphologically diverse group including flagellated planktonic forms such as *Chlamydomonas*, and colonial, mucus-covered forms (*Sphaerocystis*, Fig. 3.7), as well as large, mainly benthic filamentous forms (*Spirogyra*). Many green algae have a rigid external wall made of cellulose and have bright green chloroplasts including chlorophyll *b* as a major component.

Golden-brown algae   Chrysophytes have a golden-brown colour because they have carotenoid pigments in addition to chlorophyll. Most chrysophytes

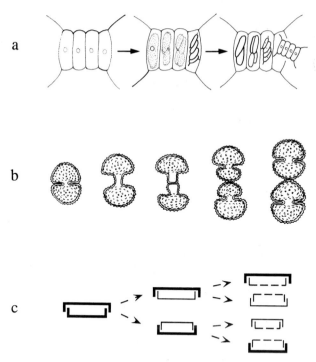

Fig. 3.9   Examples of asexual reproduction in algae. (a) Autospore formation in *Scenedesmus* sp. (class:Chlorophyceae), where new cells are formed within the old cells; (b) cell division shown by *Cosmarium* sp. (class: Conjugatophyceae); (c) the division of diatoms starts when the cell halves loosen from each other and new shell halves develop. Since the old halves always constitute the 'lid', the mean cell size in a diatom population decreases over time. When a critically small size is reached, cell division ends and auxospores develop. From these spores, cells with the original size 'hatch'.

are unicellular and some are covered by delicate siliceous or calcareous plates. Some colonial forms occur, including the common and widespread *Dinobryon* (Fig. 3.6). The *Dinobryon* cells live within vase-shaped loricas, which are attached to each other giving the characteristic fan shape to the colony. Chrysophytes form thick-walled resting spores of silica which can stay dormant in the sediments during long periods.

Diatoms   The majority of the diatoms (Bacillariophyceae) are periphytic (i.e. associated with a substrate surface), but they are also an important component of the phytoplankton. They are often dominant at high pH, irrespective of the nutrient status of the lake or pond (Fig. 3.8). Diatoms occur in centric and pennate forms (Fig. 3.6). The characteristic cell wall of diatoms is silicified with sculpted ridges and grooves, and consists of two parts, one fitting over the other as a lid. When the cell divides, the two parts are separated and new valves are formed on the freshly divided surfaces (Fig. 3.9). Since the parent valve always constitutes the 'lid' of the daughter cells, one of the cells is of identical size to the mother cell, whereas the other daughter cell will be smaller, leading to a progressive decrease in the average cell size of a diatom population as time goes by. However, when a minimum critical size has been reached, so-called **auxospores** may be formed after sexual reproduction. The enlarged auxospore cells start to divide and produce new cells that are of the original, large size. Thereafter, asexual reproduction starts again.

Dinoflagellates Dinoflagellates are unicellular, flagellated, and usually motile algae. Some are 'naked' (*Gymnodinium*, Fig. 3.6), whereas others have thick cell walls (*Ceratium*, Fig. 3.7; *Peridinium*, Fig. 3.6). They have two flagellae: one that goes horizontally around the body, usually in a groove, the other passing longitudinally and hanging out behind the algae. Dinoflagellates form resting cysts at the sediment surface which hatch when conditions are favourable for that specific species.

Euglenoids (Euglenophyta) Euglenoids are generally associated with high amounts of dissolved organic matter (i.e. are often abundant in humic lakes or close to the sediment surface). They have two flagellae (although only one is visible) arising from an anterior flagellar pocket, making the algae motile. Many euglenoids are facultative heterotrophs and therefore have an ingestion apparatus (i.e. they do not entirely rely on photosynthesis as their energy source), but may also consume bacteria and detritus. Common and well-known representatives are *Phacus* (Fig. 3.6) and *Euglena* of which the latter often becomes dominant in polluted water.

Periphytic algae

Periphytic algae, or 'Aufwuchs', refer to the microalgae attached to a substrate. The substrate may be a plant and the algal assemblage is then called **epiphytic**; if the substrate happens to be sand or mud, the assemblage is called **epipsammic** and **epipelic**, respectively. Although this classification may occasionally be useful, the algae themselves are seldom substrate-specific and the composition of the assemblage is mainly determined by other factors than the substrate material, such as current velocity, nutrient availability, and pH of the surrounding water. Moreover, when algae grow on artificial substrates, such as plastic—which they do—what shall we then call them: epiplastic algae? Owing to these problems, we will here choose the pragmatic solution and refer to microalgae growing on a substrate as 'periphytic' algae, irrespective of whether the substrate is a plant or an old bike.

A periphytic algal assemblage may be seen with the naked eye as a greenish or brownish layer covering most surfaces in lakes and ponds. Many of the algal species found in the phytoplankton community are also found among the periphyton, especially diatoms and some cyanobacteria (e.g. *Microcystis*, Fig. 3.6). Periphytic algae are important because they produce oxygen at the sediment surface, which, as we have seen in Chapter 2

(*The abiotic frame*), reduces the phosphorus transport from sediment to water, and also allows other, oxygen-demanding, organisms to live near the sediment surface. Moreover, periphytic algae serve as a major food source for many invertebrates, including chironomids and snails. They become especially important in clear water, such as in lakes and ponds at high latitude or altitude, where nutrient concentrations in the water column are low and where light reaches a large area of the sediment surface. In such lakes, periphytic algae may account for up to 80% of the primary production!

When a pristine surface area occurs (e.g. when a stone is thrown into a lake), it is first colonized by bacteria and then by diatoms. These pioneers have a large part of their surface in contact with the substrate to which they attach by secreting a sticky slime, making them resistant to turbulence as well as grazing. In the shelter of these pioneers, stalked species and rosette forms occur (Fig. 3.10), followed by filamentous forms which eventually become dominant by being superior competitors for light, a limiting resource. It is worth noting that this microtopography has a parallel in an ordinary terrestrial forest with grass, bushes, and trees, and is created by the same striving to optimize photosynthesis. However, the final stages in this ideal succession of periphytic algae are more vulnerable to animal grazing than the pioneer stage, leading to a reversal of the succession at high grazing pressure.

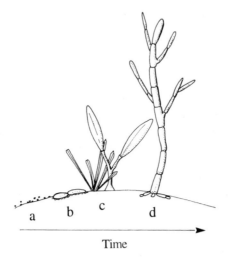

Time

Fig. 3.10 The succession pattern of periphytic algae on a 'clean' substrate. After initial rapid colonization by bacteria (a); small, flat cells invade which have a large part of the cell surface attached to the substrate, such as small diatoms (b). Thereafter, standing or stalked forms become abundant, and eventually filamentous forms (d) will occur. Note that this colonizing pattern is similar to that of a terrestrial forest with grass, bushes, and trees!

Metazoa—animals with differentiated cells

Freshwater sponges

Freshwater sponges are commonly found growing on solid substrates such as stones, macrophytes, and branches of fallen trees. Here, they form a greenish or yellowish crust, sometimes with finger-like projections (Fig. 3.11). Variation in growth form can be substantial even within a species and depends on environmental factors such as wave action and light availability. **Feeding:** sponges are filter-feeders that are capable of ingesting particles ranging in size from large algae to small bacterial cells. A large amount of water can be filtered by sponges and in extreme cases it has been estimated that the sponge population of a pond has the capacity to cycle a water volume equivalent to the volume of the whole pond over just one week (Frost *et al.* 1982). In addition, sponges may contain considerable numbers of symbiotic algae (zoochlorellae) giving it a bright green appearance (see Chapter 4, *Biotics*). **Life cycle:** sponges reproduce both through asexual and sexual processes. Asexual reproduction occurs when fragments of the cell are broken off and develop into new, active sponges. In temperate lakes most sponges form a dormant stage (*gemmule*) for overwintering. During the period of active growth sponges may also reproduce sexually.

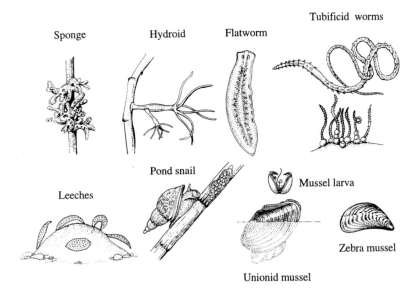

Fig. 3.11   Examples of benthic macroinvertebrates common in lakes and ponds. The upper row shows the pond sponge *Spongilla lacustris*, a hydroid (*Hydra* sp.), a flatworm (*Planaria* sp.), and an individual and a colony of tubificid worms (*Tubifex* sp.). The lower row shows leeches (*Glossiphonia complanata*) moving and sitting adhered to a stone, a pond snail (*Lymnaea stagnalis*) with an egg capsule, a unionid mussel (*Anodonta cygnea*) with a larva, and a zebra mussel (*D. polymorpha*).

Male sponges release sperm that are taken up by female sponges through their water currents. Larvae developed within the female sponge are released, disperse by swimming, and finally settle on a suitable substrate where they metamorphose and develop into a new sponge.

A large number of small invertebrates, such as protozoans, oligochaetes, nematodes, mites, and insect larvae, live within or on sponges. Some associations are obligate and involve species that are highly specialized consumers of sponge. However, predation by these smaller invertebrates does not seem to be lethal to the sponge. Sponges are protected against predation from larger predators by having a mineral skeleton consisting of needle-shaped silica formations (*spicules*) and by being toxic.

Hydroids

Hydroids are members of a group of animals (Coelenterates) that are richly represented in marine habitats, including organisms such as jellyfish, corals, and sea-firs, whereas in freshwaters only a few species occur. All hydroids have a simple body plan with a central cavity surrounded by a wall with only two cell layers and they grow attached to hard surfaces (Fig. 3.11). **Feeding:** prey organisms (small crustaceans, insect larvae, and worms) are captured with the aid of stinging cells, *cnidoblasts*. The cnidoblasts shoot out a sticky thread that entangles the prey which is then moved to the mouth opening with tentacles. The stinging cells are also an effective defence against predators, although some flatworms, crayfish, and a ciliate are known to feed on hydroids. Some hydroids also eject a neurotoxin that paralyses the prey. Green hydroids contain a symbiotic algae. **Life cycle:** during periods of optimal environmental conditions, hydroids reproduce asexually by forming buds that separate from the parent. Under more severe environmental conditions, hydroids reproduce sexually and the fertilized egg can enter a resting stage which develops only when conditions are favourable.

Flatworms

Flatworms are divided into three groups: Turbellaria, Trematoda (flukes), and Cestoda (tapeworms). **Feeding:** turbellarian flatworms are free-living carnivores, whereas trematodes and cestodes are parasitic, often with a freshwater organism as a host or intermediate host in their complex life histories. Most large freshwater turbellarians are triclads, a group characterized by having a flattened, unsegmented body with large numbers of cilia with which they glide over the substrate (Fig. 3.11). Triclads feed on small invertebrates, such as insect larvae and crustaceans, which they first wrap up in slime. **Life cycle:** triclads are hermaphrodites and deposit egg cocoons on stones or water plants. Some species may multiply by dividing into two parts that develop into two new animals. A remarkable phenomenon in triclads is their ability to regenerate into a complete new animal from a small piece, which has been accidentally cut-off its body. Predation on triclads is suggested to be minor, although some fish do include them in their diet.

Worms

Annelid worms have an elongated, cylindrical, or sometimes flattened, soft body which is divided into segments. They are hermaphrodites and eggs are deposited in a cocoon. Two major groups of worms inhabit freshwaters: *oligochaetes* and *leeches*. In oligochaetes, each segment has four bundles of bristles (chaetae) which are used to grip the sides of their burrows. The number and shape of the chaetae are used for species identification. As oligochaetes and leeches differ so much in their ecology we treat them separately.

Oligochaetes

Oligochaetes commonly construct tube-like burrows in soft sediments, with the tails extending up into the water column. The tail functions as a gill and is moved around to increase oxygen uptake (Fig. 3.11). Tubificids, which are often found at high densities in habitats enriched with organic pollutants, have physiological adaptations (haemoglobin) to survive periods of anoxia. **Feeding:** most oligochaetes feed by ingesting the soft sediments that they live in and, as is the case for most detritivores, bacteria, and other micro-organisms that have colonized the sediment are the most important source of nourishment. Some naidid worms graze on epiphytic algae. A number of predators, both invertebrate and vertebrate, feed on oligochaetes, including, for example, predatory insect larvae, leeches, and benthivorous fish.

Leeches

Leeches can be distinguished from other worms and from flatworms by having two suckers, one at each end of the ventral side of the body. They use the suckers to attach to solid substrates and to their prey. The position and number of eyes are important diagnostic features for species determination. One group of leeches consists of species that have a proboscis (i.e. elongated mouthparts used to penetrate the tissues of the prey organism and to suck out its body fluids). Another group of leeches are more parasitic, feeding on the blood of their hosts. They have toothed jaws that rupture the body surface of the prey and their saliva contains a substance that prevents blood from coagulating. One large species, the medical leech *Hirudo medicinalis*, was commonly cultivated and used for bloodletting during the seventeenth to nineteenth centuries, a practice that is even used today (e.g. in complicated hand surgery). **Feeding:** sanguivorous (blood-feeding) leeches may take a meal and then digest it for weeks or months until the next meal. Some leeches have neither toothed jaws nor a proboscis, but swallow their prey whole. The leech *Erpobdella octoculata*, for example, feeds on chironomids, blackfly larvae, caddisfly larvae, and oligochaetes and if you collect this species you will quite commonly find individuals with a chironomid sticking out of their mouths. **Life cycle:** leeches deposit their eggs in cocoons and, further, glossiphonids brood their cocoons and carry the newly hatched juvenile leeches on the underside of the body. Fish, birds, and invertebrate predators feed on leeches.

Snails

The most characteristic feature of freshwater snails is of course the shell. This is generally coiled into a spiral, either in a flat spiral as in the ramshorn snail *Planorbis corneus*, or pointed as in the pond snail *Lymnaea stagnalis* (Fig. 3.11). In addition, a few freshwater limpets (e.g. *Ancylus fluviatilis*) have a conical shell. The shell consists of an outer layer (periostracum) of organic material and a thick inner layer mainly composed of crystalline calcium carbonate. The unsegmented, soft body can be withdrawn totally into the shell. Snails have a distinct head with tentacles with eyes at their base. They glide over the substrate surface with their muscular foot. To facilitate movement, the foot is covered in mucus and you can often see a mucus trail where snails have moved. One group of snails, the *prosobranchs*, breathe by moving water over a gill. *Pulmonates*, on the other hand, breathe by going to the water surface and bringing air into a richly vascularized cavity (the 'lung'). Some pulmonates, however, fill the lung with water and rely on diffusion of dissolved oxygen. Prosobranchs are further characterized by having an operculum, a horny plate which is attached to the foot and closes the shell opening as the snail withdraws, protecting the animal from shell-invading predators. **Feeding:** snails feed with a rasping tongue, the radula, which has transverse rows of chitinous teeth and is moved backwards and forwards to remove periphyton and detritus from surfaces. The main food sources of freshwater snails are periphytic algae and detritus. **Life cycle:** pulmonate snails are hermaphrodites and although most eggs are cross-fertilized they also have the ability to fertilize their own eggs. Most prosobranchs have separate sexes. Eggs are laid in batches within a gelatinous, transparent capsule and the egg masses are attached to stones, pond weeds, or other solid substrates, sometimes even on the shell of other snails. Some prosobranch species (e.g. *Viviparus viviparus*) deposit their eggs in a fold within the body where they hatch and are then born as fully developed individuals. Most pulmonates have an annual life cycle with overwintering individuals that reproduce in spring and then die off, whereas prosobranchs usually live and reproduce for several years.

Predators feeding on snails can be divided into two groups: shell-crushers, such as the pumpkinseed sunfish and crayfish which are specialized molluscivores, or shell-invaders, such as leeches, flatworms, and different aquatic insect larvae.

Mussels

The body of mussels is completely enclosed within two shell valves, and they are therefore called Bivalvia. The valves are joined by an elastic hinge ligament that forces the shell valves to open when the two well-developed shell adductor muscles are relaxed. The shell is secreted at the edge of the mantle, which encloses the body. Growth annuli are formed on the shell because of seasonal differences in shell growth (Fig. 3.11). Bivalves have greatly enlarged gills with long, thin filaments that, besides being used for respiration, function as a filtering apparatus. The large, muscular foot is used for burrowing and locomotion. The zebra mussel, *Dreissena polymorpha* (Fig. 3.11), attaches to solid substrates with proteinaceous byssus threads and may form aggregates with many layers. **Feeding:** mussels feed on particulate organic matter in the water, such as phytoplankton, bacteria, and detritus. The ciliated gills create a water current that draws water into the shell through the lower of two siphons. Particles are removed by the filtering gill and food particles are transported to the mouth opening, whereas non-food particles are bound in mucus and excreted as pseudofaeces. Dense populations of mussels may filter a substantial portion of the water mass and clear the water of suspended particles. Further, excretion of faeces and pseudofaeces may increase the organic content of sediments locally and improve the environment for other sediment-living detritivores. **Life cycle:** the life history strategies and reproduction differ quite markedly among groups of mussels. The large (10–15 cm) unionid mussels, such as *Anodonta*, *Unio*, and *Elliptio*, release sperm into the water which are then taken up in the inhalant water of other individuals. The fertilized eggs are retained in the outer gills where they develop into larvae (glochidia) after a period of up to a year. Following release from the female the larva has to attach to a fish within a few days otherwise it dies. Some unionid larvae may attach to a number of fish species, whereas others have specialized on a single host fish species. Once attached to a fish the larva encysts and metamorphoses into a mussel, which then falls to the bottom, ideally at a place with soft sediments. The unionids release a very large number (0.2–17 million per female and breeding season) of very small (0.05–0.4 mm) larvae and a very small proportion survive to the adult stage. A completely different life history strategy is shown by the sphaeriid clams (*Sphaerium*, *Pisidium*, *Musculium*). These small (2–20 mm) bivalves brood their young in specialized brood chambers and do not release them until they are fully developed. A third mode of reproduction is shown by the zebra mussel. Here, the sperm and eggs are both released into the water where the eggs are fertilized and develop into free-swimming larvae. After about 10 days the larva settles on a hard substrate and develops into a zebra mussel.

Few predators can feed efficiently on large unionids allowing a high probability of survival once the mussels have reached adult size. However, mammal predators such as muskrats, otters, and racoons may feed on even the large unionids. Sphaeriids and juvenile unionids are often consumed by molluscivorous fish, shorebirds, ducks, and crayfish.

Rotifers

One common group of Metazoa is rotifers, which are small, generally between 0.1 and 1 mm long. They play an important role in aquatic systems mainly because of their enormous reproductive potential, which allows them to occupy short temporal niches: they may occur in numbers as high as 20 000 animals l^{-1} in natural waters. Most rotifers are solitary, but some form permanent colonies, such as *Conochilus* sp. (Fig. 3.12), which can be seen with the naked eye as dots in the water. As for many crustacean zooplankton, rotifers migrate vertically through the water column, although their migrations are less dramatic and they travel a distance of only a few metres per day. All rotifers have two distinctive morphological

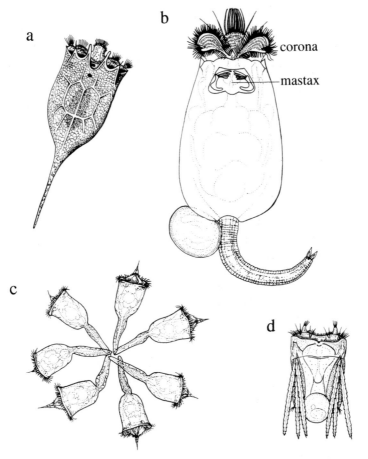

Fig. 3.12   Rotifers. (a) The rotifer *Keratella cochlearis*, common in many lakes; (b) a schematic drawing showing the two characteristic morphological structures for rotifers:the ciliated *corona*, and the *mastax* used for grinding food particles. The size of both animals is about 100 μm; (c) the colonial rotifer *Conochilus*. This specific colony contains seven animals and each of them is about 300 μm long, making it clearly visible to the naked eye; (d) *Polyarthra* sp.

features: a *corona* and a *mastax*. The corona is a ciliated region used in locomotion and food gathering (Fig. 3.12) and allows rotifers to swim in their characteristic rotating mode, which has given them their name (from *rota*, 'wheel', and *ferian*, to transport). The other feature is a muscular pharynx, the mastax, where two grinding plates crush the food. **Life cycle:** rotifers are generally parthenogenetic (i.e. the egg develops without fertilization into a new individual identical to the mother). Species that are not entirely parthenogenetic produce males only during a short period of the year. Fertilized eggs are developed into thick-walled resting eggs. **Feeding:** most rotifers are filter-feeders and may be characterized as generalist feeders, eating bacteria, algae, and small ciliates. Filter-feeding rotifers are able to eat particles up to about 18 μm, which is about the size of a small alga. Besides locomotion, the corona is used to create a water current passing by the mouth of the rotifer—a feeding current. Once in the gut, the food particle reaches the mastax and is crushed (Fig. 3.12). Although rotifers are small animals, their filtering capacity is enormous and they can filter up to 1000 times their own body volume each hour! This means that rotifers incorporate a lot of food particles into their body mass, energy that can be used by organisms further up in the food web. Predatory rotifers, such as *Asplanchna*, mainly eat other rotifers and ciliates, but are also interested in vegetables, such as algae.

Crustaceans

Crustacea include a large number of species that have successfully inhabited freshwater systems. They come in many different shapes and sizes, having evolved to exploit different habitats and resources. Some cladoceran species even change in shape over the growing season; **cyclomorphosis** (Fig. 3.13). Microscopic crustaceans (cladocerans and copepods) are the main component of the zooplankton in most pelagic food webs, whereas larger, benthic isopods and amphipods are important detritivores and herbivores in littoral habitats. The largest freshwater crustaceans are the omnivorous crayfish, commonly used as a food resource by man. Although there are appreciable differences between the different groups, they also have a lot of features in common. Here, we begin with an overview of aspects of morphology, physiology, and ecology that are shared by all groups and then we will treat the most important groups and their specifics in greater detail, starting with the zooplankton crustaceans and ending with crayfish.

Crustacea belong to the phylum Arthropoda, together with Insecta and Arachnida (spiders and mites). They are characterized by having segmented bodies that can be divided into three regions: the head, thorax, and abdomen. In some taxa the first two regions are fused, as in crayfish where the head and thorax are combined and covered by a large shield, the

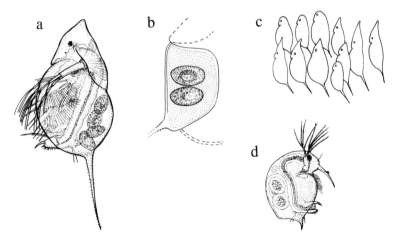

Fig. 3.13   (a) A fully grown *Daphnia* individual, one of the most efficient planktonic herbivores (sometimes called 'the cattle of lakes and ponds'), with four visible parthenogenetic eggs; (b) details of two resting eggs contained in the ephippium formed after sexual reproduction; (c) size and shape variability within a *Daphnia* species found over a season ('cyclomorphosis'); (d) *Bosmina*, a smaller cladoceran zooplankton species.

carapace. The crustacean body is covered by a hard exoskeleton consisting of chitin and calcium carbonate. The hard segments are joined with a thinner more flexible cartilage which enables movement. Muscles are attached to the hard exoskeleton which provides protection against drying out and predators. However, the exoskeleton creates difficulties for a growing animal due to its inflexibility—the animal simply outgrows it. Therefore, crustaceans have to shed periodically and develop a new, larger exoskeleton that allows growth (i.e. the animal moults). During these periods when the old exoskeleton is shed and the new one is still soft, the animal is very vulnerable to predators and thus often becomes inactive and hides in shelter. Crustaceans have two antennae and all or most segments of the exoskeleton have a pair of additional appendages. These are specialized for different functions in different species, including feeding, walking, swimming, respiration, reproduction, or defence. Crustaceans either use gills for breathing or depend on oxygen transport directly across the body surface (smaller crustaceans). There are both males and females but in some species males are only present during short periods of the year. The female employs a simple form of parental care in that she retains the eggs in a brood chamber or attaches them to her body. The eggs hatch to a larval stage, the nauplius, which remains in the egg in most species but is free-swimming in some (e.g. copepods). The nauplius metamorphoses to a juvenile that looks identical to the adult. Some species retain the small juveniles in the brood pouch for some weeks before they are released into the water.

Crustacean zooplankton

There are two important groups of crustacean zooplankton: cladocerans and copepods. Both are found in most lakes and ponds in the world, both in pelagic and benthic habitats.

Cladocerans Cladocerans are small, generally transparent crustaceans, which are commonly called 'water fleas' due to their shape and jerky swimming mode. Their body is disc-shaped and covered with a carapace attached only at the back of the neck, protecting the body like an overcoat. Attached just below the head are the major swimming organs, which are actually the second pair of antennae, rebuilt as paddles. The most well-known genus of cladocerans is *Daphnia* (Fig. 3.13), which is the major herbivore in many lakes and ponds and could be called the 'grazing cattle' of lakes and ponds. Other well-known cladoceran genera are the pelagic *Bosmina* (Fig. 3.13) and the mainly benthic *Chydorus*. **Feeding:** many cladocerans, including most *Daphnia* and *Bosmina* species, are pelagic filter-feeders mainly eating algae and, to some extent bacteria. *Daphnia* are generalist filter-feeders with a broad food-size spectrum and their gut content can serve as a good qualitative representation of the species composition in the algal and bacterial assemblages of the surrounding water. A large acceptance range in food particle size makes the grazer less vulnerable to fluctuations in abundance of specific size classes of food (such as small algae), since it can easily take advantage of alternative size classes (such as bacteria or larger algae). Due to its rapid filtration rate and large acceptance range with respect to food particle size, *Daphnia* constitutes a serious threat to algae. Other cladocerans, such as *Chydorus*, are benthic, typically feeding by crawling along surfaces, scraping, and filtering food particles. The large cladoceran *Leptodora*, on the other hand, is mainly predatory, grasping large prey, such as rotifers, ciliates, and even copepods. **Life cycle:** cladocerans are able to reproduce parthenogenetically. The space between the carapace (the 'overcoat') and the body is used by the females as a brood chamber. Here, eggs develop into embryos, and eventually assume the general shape of adults, which are then released into the water. Parthenogenetic reproduction continues until unfavourable conditions arise, then some eggs develop into males and others into haploid eggs that need fertilization. Induction of male production is enhanced by high population density or rapid reduction in food supply. Following fertilization, the walls of the brood chamber of the female thicken, forming an **ephippium**, or resting egg, which is easily recognized as a dark saddle-shaped back-pack (Fig. 3.13). The ephippium can withstand severe conditions, including freezing or drying and may be dispersed great distances by, for example, birds. In favourable conditions, the ephippium hatches into new parthenogenetic females that rapidly start to reproduce.

Cladoceran population dynamics are characterized by 100-fold annual variations in population size. In combination with high grazing efficiency,

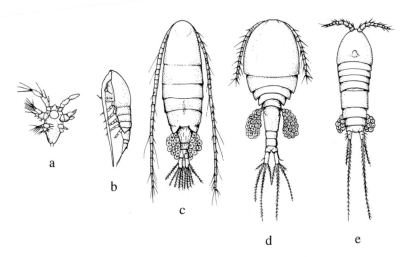

Fig. 3.14   Copepods. (a) A nauplius larva hatches from a copepod egg; (b) after five to six stages of nauplius larvae, a copepodite is formed, which looks almost like a fully grown copepod. Five to six stages of copepodites have to develop before the adult copepod appears. Three types of copepods occur: (c) *calanoid copepods*, characterized by long antennae, a slender body, and only one egg sac; (d) *cyclopoid copepods*, characterized by a more rounded body shape, short antennae, and two egg sacs; and (e) *harpacticoids copepods* characterized by an almost cylindrical body shape, very short antennae, and two egg sacs.

such considerable changes in abundance strongly influence the biomass dynamics of the algal assemblage. Moreover, cladocerans are also important as a major food source for many fish species.

Copepods   Copepods are present in almost all freshwaters, from high altitude melt water lakes to lowland ponds. The free-swimming copepods can be divided into *calanoids, cyclopoids*, and *harpacticoids* (Fig. 3.14). Copepods have, in contrast to the disc-like shape of cladocerans, a somewhat cylindrical and segmented body. They generally range in size from less than 0.5 to 2 mm, although some species may reach lengths of more than 5 mm.

Like cladocerans, copepods sometimes form swarms which may, just as for fish larvae, be a way to reduce predation. Swarms may also be aimed at facilitating encounters between sexes, just like the swarming behaviour of humans in urban habitats! Calanoid copepods have antennae that are usually longer than the body, whereas cyclopoids have short, and harpacticoids even shorter antennae. Harpacticoids also have a more cylindrical body shape than the other two groups (Fig. 3.14). **Life cycle:** copepods generally have sexual reproduction, and when eggs are fertilized, they are extruded in one (most calanoids and harpacticoids) or two (cyclopoids) egg sacs. In suboptimal conditions, such as food shortage and high predation pressure, copepods may produce resting eggs with thicker walls able to survive long periods in the sediments (see Chapter 4, *Biotics*). After hatching, five to six naupliar larval stages follow, which in turn are followed by five to six copepodite stages before the adult copepod appears (Fig. 3.14). Hence, the developmental period from egg to adult is much longer than in cladocerans, and takes from 10 days to a month depending on temperature. **Feeding:** copepods have a varied diet and may be herbivores, predators, detritivores, or omnivores. Most calanoids have filtering mouthparts and are herbivores, whereas harpacticoids are almost exclusively littoral and have mouthparts mainly adapted for scraping and seizing particles from surfaces. Cyclopoids prefer a diverse diet and are usually considered to be omnivorous, eating algae as well as other zooplankton. They lack filtering mechanisms and their feeding mode is raptorial (i.e. they 'pick' food particles). Cyclopoid copepods have a typical 'hop and sink' swimming mode, easily seen by the naked eye. The 'hops' are used when attacking a prey. Various invertebrate predators and fish use copepods as prey.

Mysids

The opossum shrimp (*Mysis relicta*) has a slender, shrimp-like body with elongated appendages that are designed for fast, active swimming (Fig. 3.15). Mysids have no gills but take up oxygen directly across the thin exoskeleton. This makes them sensitive to poor oxygen conditions and their main distribution is in northern cold, deep, oligotrophic lakes. They migrate vertically in the water column and spend the nights filter-feeding on zooplankton and phytoplankton in shallow waters. Days are spent in the dark, deep water close to the lake bottom where they are less at risk of predation by visually hunting fish predators. *Mysis* has been introduced to numerous lakes with the intention of increasing the food base for fish. Instead, they have become a serious problem in many lakes, competing with juvenile fish for zooplankton or even preying on newly hatched fish larvae.

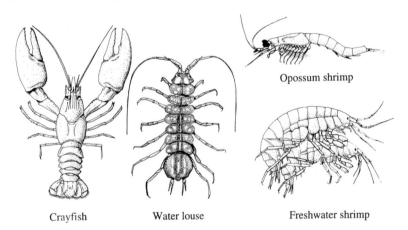

Opossum shrimp

Crayfish   Water louse   Freshwater shrimp

Fig. 3.15 Examples of crustaceans common in ponds and lakes. The noble crayfish (*A. astacus*), the water louse (*A. aquaticus*), the opossum shrimp (*M. relicta*), and a male freshwater shrimp (*Gammarus pulex*) carrying a female before mating.

Isopods

The water louse (*Asellus aquaticus*) is a typical isopod with a dorsoventrally flattened body (Fig. 3.15). It is common in the weed beds of many ponds and lakes, where it feeds on periphyton, as well as on bacteria and fungi associated with detritus. The water louse reproduces in spring and the female carries the eggs in a brood chamber, the *marsupium*, on the underside of the body. The juveniles are retained in the marsupium some weeks after hatching. Isopods are an important prey organism for many freshwater predators, both invertebrates, such as dragonfly larvae, and benthivorous fishes.

Amphipods

Freshwater shrimps, such as *Gammarus* spp. have a slender curved body shape (Fig. 3.15). They swim sideways using the legs, but may also jump by stretching the body. **Feeding:** gammarids are opportunistic foragers, feeding on whatever is available. The main part of the diet consists of periphyton, leaves, and detritus with associated bacteria, but they also feed on dead animals and may also be predatory or even cannibalistic. **Life cycle:** the eggs and juveniles are carried around in the marsupium, as in isopods. *Gammarus* may become very abundant in ponds without fish or in habitats with a high structural complexity, such as in dense macrophyte stands. However, in the presence of fish they decrease their activity and take cover under stones, etc. Fish and large predatory invertebrates are important predators on amphipods.

Decapod crayfish

Crayfish look like small lobsters. They are the largest, most long-lived crustaceans of northern temperate ponds and lakes. They have five pairs of legs of which the first pair form large, powerful pincers used for crushing food and as weapons against predators and in territorial fights with conspecifics (Fig. 3.15). Crayfish move slowly forwards by using the four pairs of legs, but when aroused by danger they may rush backwards by flexing the tail underneath the body. **Feeding:** crayfish are generalists, opportunistic feeders, and include periphyton, macrophytes, detritus, as well as invertebrates and decaying animals in their diet. **Life cycle:** typically, reproduction occurs in late autumn and the female then carries the eggs on her abdomen until late next spring. The juvenile crayfish stay attached to the female for a few weeks and then leave after their second moult. Fish are the most important crayfish predator. In Europe, perch and eel are known to consume crayfish, and in North America the diet of small-mouth bass consists to a large extent of different crayfish species. Crayfish fisheries are of great economic importance in many countries. In the twentieth century this industry has suffered greatly in Europe through a dramatic reduction of the stocks of the noble crayfish (*Astacus astacus*) due to crayfish plague (a fungus, *Aphanomyces astaci*). Because of this plague, many European lakes and ponds have been stocked with exotic crayfish species, such as the North American signal crayfish (*Pacifastacus leniusculus*), which is immune to the fungus. However, because the signal crayfish may carry the plague, the noble crayfish are less likely to recolonize waters where signal crayfish have been introduced, and are now becoming exceedingly rare.

Insects

An estimate of the total number of aquatic insect species shows that there are more than 45 000 species world-wide (Hutchinson 1993), and most of them belong to the order Diptera (>20 000). Just as for crustaceans, the insect body can be divided into three regions, the head, thorax, and abdomen. The body is covered by a hard chitinous cover which has to be shed as the insect grows. **Life cycle:** the life cycles of aquatic insects can be quite complex, but some major patterns can be recognized. First, most aquatic insects do not spend all life stages in the water and the adults are usually terrestrial and have completely different morphology than their aquatic juvenile stage (Fig. 3.16). Aquatic hemipterans and some of the beetles, however, live their entire life in water. Mayflies (Ephemeroptera), dragonflies (Odonata), and true bugs (Hemiptera) have three developmental stages—*egg, nymph, adult*—with no complete metamorphosis. The nymphs more or less resemble the adults and the wings are developed as external wing pads. Caddisflies (Trichoptera), alderflies (Megaloptera), beetles (Coleoptera), and flies (Diptera) have four life stages—*egg, larva, pupa,* and *adult.* The larva bears no resemblance to the adult; legs are often short or absent and wings develop internally. These insects metamorphose during the pupal stage. Females that reproduce during the terrestrial stage may deposit their eggs at random when flying over the water surface, but

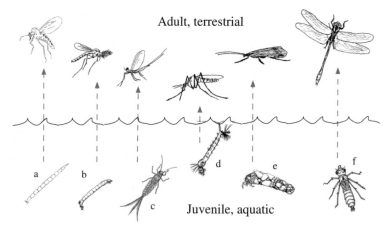

Fig. 3.16   The figure illustrates the huge differences in morphology between the juvenile (larvae or nymphs), aquatic and the adult, terrestrial forms of some common insects. (a) Biting midges (e.g. *Ceratopogonidae*); (b) non-biting midges, *Chironomidae*; (c) mayflies (e.g. *Cloeon*); (d) mosquitoes (e.g. *Culex*); (e) caddis flies; (f) dragonflies. Note that the images are not drawn to scale.

in many species eggs are attached to, for example, plants above or below the water surface. The majority of insects are benthic, living in or on the sediment surface or on the macrophytes in the littoral zone. Notable exceptions are the planktonic phantom midge (*Chaoborus*) and insects living on the water surface, such as water-striders and whirligig beetles. Aquatic insects commonly reach high population densities and because they are extensively consumed by fish, they provide an important link between primary producers/detritus and the top predators in aquatic food webs. Below, we describe some of the more important aquatic insects in more detail.

Mayflies

Mayfly nymphs (Ephemeroptera) can be readily distinguished from other aquatic insects by their three long tail filaments and rows of gills on the sides of the abdomen (Fig. 3.17). Some species swim with undulatory movements (e.g. *Leptophlebia*, *Cloëon*), whereas others are sluggish crawlers on the sediment or macrophyte surfaces (*Caenis*, *Ephemerella*). **Feeding:** most mayflies feed on periphytic algae or detritus, but a few species may also be carnivorous. Large ephemerid mayflies (*Ephemera*, *Hexagenia*) burrow into soft sediments and create a water current through a U-shaped burrow. These larvae are primarily filter-feeders and feed on particles transported through the burrow by the water current. **Life cycle:** the fully grown mayfly nymph crawls or swims to the water surface and moults into a subadult stage, the subimago, which flies to and settles on nearby vegetation where it moults to the final, adult stage within a day (Fig. 3.16). This subimago stage is only found in mayflies. The adult mayfly does not feed

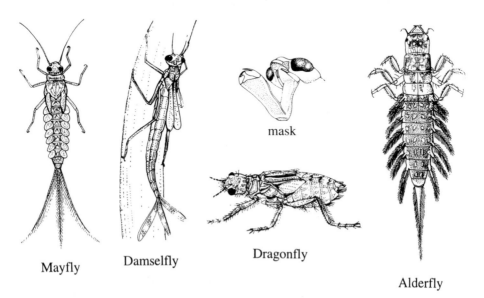

mask

Mayfly   Damselfly   Dragonfly   Alderfly

Fig. 3.17   Examples of benthic freshwater insects: a mayfly (Ephemeroptera; *Cloeon* sp.), a damselfly, a dragonfly with the feeding mask shown above, and an alderfly (*Sialis lutaria*).

and after reproduction it usually dies within a few days. The emergence of mayflies is often highly synchronized, with millions of adults emerging from a lake within a few days. This synchronization of emergence has been interpreted as a defence against predation.

Damselflies and dragonflies

Odonata consists of two suborders, damselflies (Zygoptera) and dragon-flies (Anisoptera), which vary considerably in morphology (Fig. 3.17). The nymphs of damselflies are long and cylindrical and they have three caudal lamellae at the end of the body. These have many tracheae and aid in res-piration. **Life cycle:** the length varies among species but is generally long, ranging from one to three years. Dragonfly nymphs are short and stout and lack caudal lamellae. Odonate nymphs are inconspicuous in their nat-ural habitat with drab and dull colours of brown, green, and grey shades, whereas the adults often have beautiful, bright colour patterns (Fig. 3.16). Damselflies are common in littoral zones with vegetation, where they climb among the plants stalking or ambushing their prey. They swim by moving the abdomen from side to side. Dragonfly nymphs are less active and when they move they crawl slowly on the substrate. Some species bur-row in soft sediments. **Feeding:** odonates are voracious predators feeding on a wide array of invertebrate prey, but also on tadpoles and fish larvae. They capture their prey using a highly specialized mouth apparatus, the mask (Fig. 3.17). When a prey is within reach, the mask is unfolded and shot forward and the prey is seized by hooks at the end of the mask. Fish are an important predator on odonates, but in habitats that lack fish, odonates may reach high population densities and be an important struc-turing force (predation) affecting the size structure and species composi-tion of smaller invertebrates.

Alderflies

Alderflies (*Sialis*) are a small group of insects belonging to the order Megaloptera. They have an elongated, fusiform body and a heavily sclero-tized head with large, conspicuous mandibles (Fig. 3.17). *Sialis* is common in ponds and lakes and lives buried in the sediment. The larvae are predators—small larvae feed on benthic crustaceans, whereas larger larvae mainly prey on oligochaetes, chironomids, and other insect larvae. Alderflies may reach high densities, especially when the abundance of predatory fish is low, and then they exert a strong predation pressure on smaller benthic invertebrates.

Caddisflies

Caddisflies (Trichoptera) have an elongated, rather fat body where the head is heavily sclerotized, whereas the cuticle of the thorax and abdomen usually consists of thin, colourless chitin (Fig. 3.18). Characteristic of caddisflies are also the so-called pro-legs, two hooked appendages of the

Caddisfly

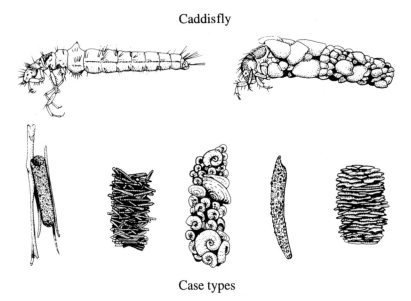

Case types

Fig. 3.18 Caddisfly larva without and with its case and examples of the elaborate caddis cases made from different materials. From the left is a sand case enlarged by sticks (*Anabolia*), a stick case, a case made of mollusc shells, a case made of sand grains, and at the far right a case made from leaf fragments.

last abdominal segment that are used to hold on to their larval case or, in free-living species, to the substrate. Most caddisflies that inhabit ponds and lakes construct a portable protective case of leaves, twigs, sand, stones, snail shells, or other materials. The shape of the case is highly variable and can often be used as a diagnostic feature for species determination. For example, *Phryganea* cuts out rectangular pieces of plant material and cements them together to a cylindrical spiral, and *Leptocerus* uses sand grains to build a narrow, conical tube (Fig. 3.18). The cases are cemented together and lined with silk produced by the salivary glands. Caseless caddis larvae are common in streams, but in standing waters only a few species lack portable cases and instead make galleries on stones or silken retreats. **Life cycle:** caddisflies have a pupal stage that they spend in a case often constructed by sealing off the old larval case at the ends and making a silken cocoon (Fig. 3.16). **Feeding:** most caddisflies are herbivores feeding on macrophyte tissue, periphytic algae as well as coarse detritus, but some species are carnivorous. They are a common prey item in fish diets.

Bugs

Water bugs belong to the order Hemiptera, which mainly contains terrestrial species. Water bugs are characterized by having mouthparts that are modified into a tube-like, piercing, and sucking proboscis and by having a pair of wings (hemelytra) that are partly hardened and protect a pair of larger

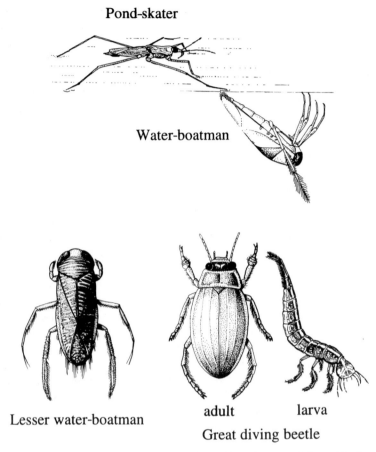

Fig. 3.19   Examples of bugs and beetles common in lakes and ponds. A pond-skater (*Gerris* sp.)
which lives on the water surface, the water-boatman (*Notonecta*) hunts for prey in the
water column but can often be found hanging on the underside of the water surface,
the lesser water-boatman (*Corixa* sp.) common in ponds without fish, and an adult and
larva of the great diving beetle (*D. marginalis*).

membranous wings. Some species lack wings completely. Water-boatmen
(*Notonecta*, Fig. 3.19), and other water bugs that actively swim, use their
hairy, flattened posterior legs as oars. Water scorpions (Nepidae) are more
sluggish, slowly moving on the bottom or among water plants. They are
easily identified by their long breathing tube at the hind end. The semi-
aquatic bugs (Gerridae) are adapted for a life on the water surface
(Fig. 3.19). Their ventral surface is covered with hairs which make them
'waterproof' and the hairs on the tarsi enable them to walk on water.
Feeding: the majority of water bugs are predators feeding mainly on
invertebrates but may also include fish and tadpoles in their diet. The proboscis is inserted into the prey and its body juices sucked out. Some lesser

water-boatmen (Corixidae) also feed on algae, detritus, and microscopic invertebrates. Pond-skaters capture invertebrates of mainly terrestrial origin that are trapped in the surface film. **Life cycle:** the aquatic hemipterans complete their entire life cycle in or on the water and only leave it during dispersal to other habitats. The nymphs look almost identical to the adults. Fish and other water bugs are the most important predators, and, in addition, cannibalism may be an important mortality source in some populations.

Beetles

Adult beetles are characterized by a heavily sclerotized, compact body. The two forewings are modified to form the hard elytra which protect the flying wings that are folded beneath them when not in use. Larvae differ considerably in shape among species, but they have a distinct, sclerotized head with well-developed mouthparts in common. Most freshwater beetles crawl on vegetation, stones, or other substrates, but some are efficient swimmers with legs that are flattened and/or have a fringe of hair. The whirligig beetles (Gyrinidae) aggregate in small schools at the water surface. When disturbed they increase their activity and eventually dive beneath the water surface. Their unusual eyes are designed for living on the water surface, and are divided into two sections, one for under- and the other for above-water vision. **Feeding:** freshwater beetles are highly variable in their feeding habits, ranging from the great diving beetle (*Dytiscus marginalis*, Fig. 3.19), which is a voracious predator both as adult and larva, feeding on invertebrates, tadpoles, and fish, to, for example, the small *Haliplus* beetles that feed on filamentous algae. **Life cycle:** In most species both adults and larvae are aquatic, and it is only the adults that leave the water during short dispersal flights. However, in some species the adult stage is terrestrial.

Flies and midges

Flies and midges belong to the order Diptera and the aquatic dipterans are a major constituent of aquatic invertebrate assemblages (Fig. 3.20). Dipterans vary greatly in morphology, but one common characteristic is the absence of legs on the segments of the thorax. Some families have a well-developed sclerotized head capsule with antennae and mouthparts, whereas others have a poorly developed or no visible head. Dipterans can be found in all types of freshwater habitats, from large lakes to small, temporary ponds and puddles. Short life cycles make them successful inhabitants of ephemeral habitats. Many species are important because the adult stage transmits diseases to humans and livestock. Also, the many biting mosquitoes, midges, and flies can be a severe nuisance in many areas (Fig. 3.16). **Feeding:** a complete range of feeding strategies can be found in dipterans, from predators to herbivores and detritivores.

The dominant family is the Chironomidae, which is by far the most common of all aquatic insects and is therefore a highly important part of

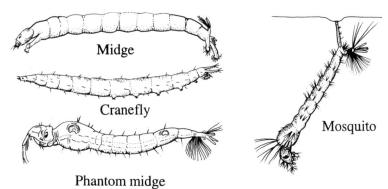

Fig. 3.20   Examples of larvae of freshwater flies and midges. Midge larva (*Chironomus* sp.), cranefly larva (*Dicranota* sp.), the transparent phantom midge (*Chaoborus*) that lives in the water column, and a mosquito larva (*Culex* sp.) common in small, temporary waters.

freshwater food webs (Fig. 3.20). Some predatory species may be free-living, but most chironomid larvae construct a loose case on the substrate surface. They feed on periphytic algae or detritus particles or filter out particles from a water current drawn through the case. Chironomids are consumed by many invertebrate predators and they make up a significant portion of the diet of many fish species.

The phantom midges (Chaoboridae, Fig. 3.20) are an important component of pelagic food webs in many ponds and lakes. The larvae have an elongated body that is almost completely transparent: only the eyes and the two air-filled hydrostatic organs can be seen. Chaoborids are predators that catch small insect larvae and zooplankton with their modified and enlarged antennae.

Fish

Everybody knows what a fish looks like, and when asked to draw a fish one generally produces something that looks like a carp. However, when looking closer at the morphology of freshwater fish, one realizes that they show an amazing diversity of forms and shapes (Fig. 3.21). The body shape of a specific fish is, of course, a result of different selection pressures, of which maximizing foraging returns is one of the most important. Fish body shapes can be divided into three categories based on the food habits of the species (e.g. see Webb 1984). First, fish that feed on prey that are widely dispersed have to spend much time and energy swimming around locating prey and should have bodies that are streamlined in order to minimize drag (e.g. salmonids or zooplanktivorous cyprinids). Second, ambush predators, such as pike, have a large caudal fin and a long, slender, flexible body that makes them specialists at acceleration, thus reducing the probability of prey

Pike

Roach

Pike perch

Tench

Large mouth
bass

Bream

Perch

Sunfish

Fig. 3.21   Examples of freshwater fish, showing differences in morphology, and position and size of fins. Fish on the left are piscivorous, whereas fish on the right are planktivorous and benthic feeders.

escape. The third category includes fish that feed on prey with low escape capacities in habitats with high structural complexity and which need a high manoeuverability. These fish (e.g. bluegill sunfish) are typically deep-bodied and have well-developed pectoral fins. These morphological categories are, of course, the extreme examples of the evolution of body shapes of highly specialized foragers. Many fish species are generalists, however, feeding on an array of prey items, or change their diet and habitat use as they grow (see below). The resulting body shape is then a compromise between different needs.

The shape and the position of the mouth, is also an adaptation to habitat use and diet. Fish that feed in open water usually have a terminal mouth, whereas fish feeding on sediment-living organisms have a mouth that is located towards the ventral side. Fish that filter-feed on small food particles commonly have numerous, long, fine gill-rakers; projections from the gill arches. Some species, such as many cyprinids, cichlids, and sunfish, have specialized pharyngeal bones that are used for crushing and chewing hard-bodied prey such as snails. In cyprinids, which may be very similar in external morphology, the shape of the pharyngeal teeth can be used for species identification. **Feeding:** freshwater fish can be found in all consumer categories, from detritivorous species and herbivores to carnivores, including primary and secondary carnivores. Herbivores may feed on microscopic algae (phytoplankton, periphyton) or on macrophytes. The carnivores include zooplanktivorous fish, fish that feed on invertebrates inhabiting the sediments or plants and other substrates in the littoral zone, and the piscivores that feed on other fish. Many species are highly specialized, such as the efficient piscivorous pike or the pumpkinseed sunfish in North America

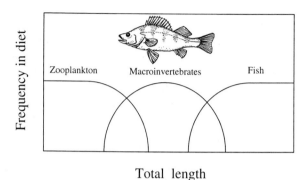

Fig. 3.22   Ontogenetic diet shift in perch. The graph shows the change in the frequency of different food items in the diet of perch as a function of increasing perch size.

which has a diet that consists of up to 90% of snails. However, other species are generalists, feeding on whatever becomes available. Fish have indeterminate growth—they continue to grow all their lives and do not stop when they have reached sexual maturity—and as they grow they often change their diet and habitat use (i.e. they go through an ontogenetic niche shift). As an example, the European perch feeds on zooplankton as a juvenile, it then turns to benthic macroinvertebrates, and when it reaches a larger size it becomes piscivorous (Fig. 3.22). These ontogenetic changes are central in understanding many of the interactions between fish, their predators, and their prey (see Chapter 5, *Food web interactions*). **Life cycle:** freshwater fish typically have separate sexes. The common mating system is ovipary, where the eggs and the sperm are ejected into the water and fertilization takes place. Fish typically invest in quantity rather than quality—females produce a large number of small eggs. In most species the parent fish do not show any parental care. The eggs may be deposited in the open water mass and left to float to the surface or sink to the bottom where they develop. Substrate spawners may deposit their eggs in long strings that get entangled in vegetation or produce adhesive eggs that stick to stones or macrophytes. However, intricate patterns of parental care are sometimes found in freshwater fish. Some fish species construct nests which may be simple depressions in the substrate (e.g. in bluegill sunfish), or elaborate nests made of weed (e.g. sticklebacks) where the eggs are deposited. The eggs, and in some cases the larvae, are cleaned and protected against predators by the parents.

Fish are probably the most important predators of other fish, both cannibalism and interspecific piscivory being important mortality sources in many fish populations. Further, large predatory invertebrates may include fish larvae in their diets (see above) and many mammals and birds are specialized fish predators. Fish are, of course, also a valued prey to humans; commercial freshwater fisheries as well as sport fishing are of large economic importance in many countries.

Amphibians

The two major groups of amphibians found in lakes and ponds are frogs and salamanders (Fig. 3.23). Some species live their whole life in freshwater whereas other species are completely terrestrial. Most amphibian species, however, have a complex life cycle where they spend their larval period in freshwater systems, undergo a dramatic metamorphosis in body morphology, leave the water and start living their adult life on land. Tadpoles, for example (Fig. 3.23), which are the aquatic stage of frogs, have no legs but a tail that enables them to swim, they breathe with gills and their mouth parts are adapted for feeding on algae. When they metamorphose, the tail is resorbed and instead they get their four legs, the gills are lost and the respiratory function is taken over by lungs, the larval mouth parts disappear and are replaced by jaws, teeth, and a tongue. The metamorphosis in salamanders is not that dramatic as they retain their general body shape and have feet and tail both as juveniles and adults. Tailfins and gills are lost, however, and lungs developed. **Feeding:** most tadpoles have a feeding apparatus that allows them to trap bacteria, phytoplankton, and other small particles suspended in the water. Many species also graze on periphytic algae and some species even have mouth parts adapted for a predatory feeding mode. Salamanders start to feed on zooplankton but as they grow they include larger invertebrates in their diet and some species even prey on tadpoles. **Life cycle:** during the breeding season, males aggregate in the littoral zone of ponds and lakes and attract females with calls that are species-specific. Females generally deposit their eggs in clusters that could be grapelike, floating rafts, or strings of eggs, although some species deposit their eggs singly. The eggs are surrounded by a gelatinous cover that gives protection against mechanical damage, desiccation, and some predators. The cover also acts has an insulator so that the heat from solar radiation absorbed by the black egg is retained in the egg cluster and speeds up development. The tadpoles that hatch from the eggs live in the pond typically 1–3 months before

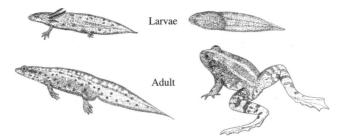

Larvae

Adult

Fig. 3.23   Examples of amphibians common in lakes and ponds. Larval salamanders (top left) and tadpoles (top right) are aquatic, whereas adult salamanders and frogs (bottom) spend a major part of their time on land.

metamorphosis. Intraspecific and interspecific competition for food in tadpoles affects the time to metamorphose and size at metamorphosis. At high tadpole densities tadpoles metamorphose later and at a smaller size, which reduce lifetime fitness. In some species the larvae will overwinter and metamorphose the next summer if their growth rate has been too poor.

Predation is an important mortality source in frogs and salamanders. A number of predators in lakes and ponds feed on eggs, tadpoles, and salamanders, including insect larvae (dragonflies, water-boatmen, diving beetles), fish, and birds. Adult frogs and salamanders are also victims of predation by fish and birds.

Birds

Birds are a common feature in the fauna of lakes and ponds. Some species are more or less obligately associated with the freshwater habitat, whereas others just take advantage of the resources of the pond or lake, such as warblers feeding on adult flying insects that spent their larval period in the lake. The birds most people directly associate with lakes and ponds are the wildfowl (waterfowl in American English) which are ducks, geese, and swans. But other bird taxa also use standing waters to different degrees, including divers, grebes, flamingoes, pelicans, herons, egrets, gulls, terns, moorhens, waders, kingfisher, and birds of prey such as ospreys and bald eagles. The more obligate water birds have evolved a number of adaptations to cope with the freshwater habitat. Webbed feet that are placed far back on the body are excellent for swimming and manoeuvring under water (Fig. 3.24). The terminal placement of the feet, however, results in that they have difficulties moving around on dry land. Diving ducks can compress their plumage to remove air and thereby decrease buoyancy so that they can dive deep down in the water column. The size of the heart of diving ducks differs among species by being larger in the species that dive the deepest, reflecting the high oxygen demand associated with deep-water diving. Herons, egrets, and waders have long legs that enable them to wade around in shallow waters when searching for food (Fig. 3.24). The size and shape of the bills vary in water birds in accordance to their major food source. Dabbling ducks have comb-like lamellae along their bills which they use to filter small seeds and invertebrates from water and mud. Birds that feed on fish, on the other hand, typically have long, spear-like bills. **Feeding:** Lakes and ponds provide a rich source of food for birds, including most organisms from filamentous algae to plant tubers and zooplankton to fish. Over time, birds have specialized on different kinds of food and also separated along different aspects of the habitat niche, such as water depth (Fig. 3.24). Many species, such as swans and dabbling ducks, feed mainly from the water surface where they skim seeds and invertebrates that have accumulated in the surface film or on plants and macroinvertebrates under water that they can reach without diving. The depth they can reach increase by up-ending, that is, when they put the

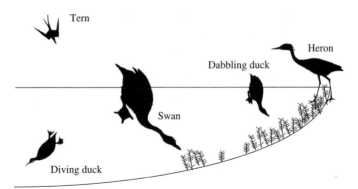

Fig. 3.24   Feeding habits of different waterfowl, including wading around in shallow waters (e.g. Herons), up-ending, that is, putting the tail in the air and have the front part of body under water (e.g. dabbling ducks and swans), diving in deep water (e.g. diving ducks and cormorants), and plunging into the water (e.g. terns).

tail in the air and have the front part of body under water (Fig. 3.24). Different species can reach different depths depending on the length of their neck and studies on waterfowl have shown that coexisting species differ in neck length and thereby separate in their resource use along the water depth gradient. Further, species foraging at the same depth may show differences in bill morphology, for example, in the number and size of straining lamellae, that allow them to specialize on different parts of the resource spectrum. Other birds submerge completely when foraging. They go under water from a position on the surface, like diving ducks and cormorants (Fig. 3.24), or plunge into the water when flying (terns, pelicans; Fig. 3.24) or from a perch at the lakeside (kingfishers). These different feeding habits allow access to different depths; that is, the waterfowl are to a large extent niche separated with respect to feeding (Fig. 3.24). A final group of water birds is represented by the geese that feed on plants in the freshwater habitat to some extent, but have their main food intake on land. Many bird species are omnivores, at least when looking over all seasons. Within seasons, diets are often more specialized. Dabbling ducks, for instance, feed on seeds and plant parts during the non-breeding season but rely heavily on invertebrate food in seasons when they need protein-rich food for growth, breeding, or moulting. **Life cycle:** birds associated with lakes and ponds exhibit the full range of mating systems even though monogamous mating, where a couple pair for the whole breeding season and sometimes for life, is predominant. They nest in places that provide maximal protection from predators, including islands, rocks, tree cavities, or on platforms built of plant material. All wildfowl (swans, geese, ducks) are *precocial*, that is, their chicks are down-covered when they hatch and are able to leave the nest to swim and forage within hours to a few days after hatching. Other species, such as the cormorant, have *altricial* young that are naked when they hatch and remain in the nest, being fed by the parents, until they are fully developed.

One special feature with birds, in comparison with other freshwater organisms, is that they can easily leave and move around among different lakes and ponds, for example, in response to changes in resource availability. Birds of course also take part in spectacular, seasonal migrations where they may migrate impressively long distances between breeding sites and overwintering habitats. These migrations allow them to exploit habitats with high resource abundances during some seasons (spring, summer) that are inhospitable due to harsh climatic conditions (ice cover, snow) during other parts of the year. Birds that aggregate at high densities at stop-over sites during migration or at their final winter quarters may affect the lake ecosystem, for example, by feeding on plants and macroinvertebrates. Defecating birds may also affect the lake by increasing nutrient input, especially if the birds have been feeding in other systems, such as is the case for geese foraging on land during day and roosting on lakes during night. Over the years, there have also been bitter conflicts between fish-eating birds and man and even though the impact of birds on fish stocks is still unclear, increasing populations of, for example, cormorants in parts of the world have put new fire to this conflict. Another important feature of birds is that they may constitute dispersal vectors for freshwater organisms, such as plankton and flightless invertebrates, that have no means of their own to move between freshwater systems but have to rely on passive transportation; for example, on feets and feathers of water-dwelling birds.

Practical experiments and observations

Selection of case material in caddisfly larvae

Background Caddisflies make their cases of different types of materials, such as leaves, sand grains, mollusc shells, or sticks (Fig. 3.18). An interesting investigation might be to discover if caddisflies select the material for their cases, or if they make them from whatever material they find.

Performance (1)Bring in some caddisflies with different types of cases (Fig. 3.18); they are generally common in the littoral zones of most lakes. Sort them into groups with similar cases and gently remove them from their cases. Put each group into different jars or aquaria where you present them with various types of material, including the material each specific group made their cases from when in the lake. Do they select the same material for their new cases? (2) Perform the experiment as in (1) but exclude the 'natural' material (i.e. the material they made their case from when in the lake). Do they make any new cases?

To discuss Which predators are the cases protecting the caddisfly from? Is any case material better than no case material?

4 Biotics: competition, herbivory, predation, parasitism, and symbiosis

Introduction

Do biotic interactions, such as competition and predation, have any effect on population dynamics or community structure? Or are natural communities structured entirely by abiotic factors such as climate? Such questions may nowadays seem naïve, but they have been the focus of lively debate throughout the history of ecology. The view that abiotic factors were the most important for structuring communities held sway for a long time, but during the 1960s and 1970s theoretical ecology developed rapidly, especially terrestrial ecology. Initially, competition was put forward as being the key structuring force in natural communities. Later, a more subtle picture emerged of how ecosystems could function, including not only abiotic factors, but biotic processes such as predation and herbivory, all of which differ temporally and spatially in significance. In freshwater habitats the importance of biotic interactions had its breakthrough even later, but today it is generally accepted that biotic interactions are essential for the dynamics and structure of freshwater communities.

The different biotic interactions between organisms can be classified according to the effects they have on each other's fitness, including: no effect (0), a negative (−), or a positive effect (+). Thus, with these classifications competitors that have a harmful effect on each other are denoted (− −). In consumer–resource interactions (predator–prey, herbivore–plant) one of the organisms benefits, while the other is harmed (+ −). The same holds for parasitism. **Amensalism** is the interaction where one species adversely affects the other, whereas the other has no effect on the first (− 0), and

commensalism is when the first has a positive effect on the other and the other has no effect on the first (+ 0). Lastly, in **mutualism** both organisms benefit from the interaction (+ +). In this chapter we will mainly focus on predation, herbivory, and competition, but also briefly discuss the impact of parasitism and mutualistic interactions, such as **symbiosis**. We will also see how freshwater organisms are affected by these biotic interactions and how they have evolved adaptations to cope with each other, such as by becoming better competitors, better predators, or by being better able to avoid being eaten. To set the scene for these interactions we will start with a short introduction to the niche concept.

The niche

The niche of an organism may be thought of as the total range of environmental variables where it can survive, grow, and reproduce. To exemplify the concept we could describe the niche of an imaginary species—the rock snail (*Snailus springsteenius*). This species can maintain a viable population in freshwaters that range in pH from, for example, 5.6 to 8.2, which is the species' niche along the pH gradient (Fig. 4.1). However, *S. springsteenius* can only thrive within this pH range if the temperature is between 5° and 20°C and if the calcium concentration is between 5 and 15 mg l^{-1}. If we add these dimensions to the graphical representation, we obtain a volume that describes the niche of the rock snail along three niche dimensions: pH, temperature, and calcium concentration (Fig. 4.1). To completely describe the niche of this species, we need to consider *n* environmental gradients and the total niche of the rock snail can then be described, although impossible to visualize, as an *n*-dimensional hypervolume (Hutchinson 1957). The **fundamental niche** describes the range of environmental conditions which a species can occupy in the absence of biotic interactions. If the

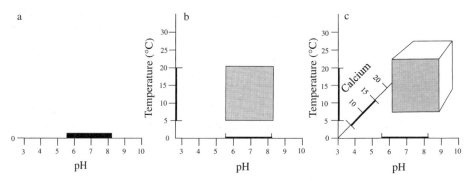

Fig. 4.1   The niche of an imaginary snail species in:(a) one dimension, along the pH axis; (b) in two dimensions, along the pH and temparature axes; and (c) in three dimensions, along the pH, temperature, and calcium axes. Modified from Begon *et al.* (1990).

fundamental niche of our imaginary snail fits into the abiotic frame of a new lake or pond that it tries to colonize, it will probably thrive and become old and wise there. However, if its fundamental niche falls too much outside the abiotic frame, the rock snail will be excluded from the list of species present in the lake.

Of course, an organism cannot exist in an environment without interacting with other organisms, such as competitors, predators, or parasites, and these interactions restrict the niche space that an organism actually occupies. This **realized niche** obviously varies between populations of the species depending on the set of organisms with which the species interacts. For example, coexisting species often have overlapping fundamental niches and if the overlap is significant with respect to a resource in limited supply, such as prey, nutrients, or nesting places, the realized niche will be very different from the fundamental niche, indicating that there is **competition** for this shared resource.

Competition

The concept of competition has been, and still is, one of the fundamentals of ecology. In the literature you can find an enormous number of studies on the effect of competition, performed in laboratory and field experiments and as observations of patterns in natural communities, on all kinds of organisms and systems, from microbes to large mammals. Given the great interest in this field there also exist a considerable number of definitions of the term 'competition'. We will not add yet another one, but adhere to the definition by Tilman (1982):

Competition is the negative effect which one organism has upon another by consuming, or controlling access to, a resource that is limited in availability.

A central concept when discussing competition is therefore **resources** which are substances or factors that result in increased growth as the availability is increased (Tilman 1982). Hence, resources include nutrients, light, food, mates, spawning grounds, and space. Only organisms that have a common resource will compete, but, and this is important, only in situations when the resource is in limited supply. If the resource is so abundant that it does not limit consumers, there will naturally not be any competitive interactions. This could happen, for example, if consumer populations are kept at low densities by other factors such as predation or abiotic disturbance events.

Competition may occur in two ways: by exploiting resources more efficiently than the competitor (**exploitation competition**), or by interfering directly with the competitor, for example, by aggressive behaviour, thereby occupying a larger part of the common resource (**interference competition**). An example of exploitation competition is given by different algal species which consume limited nutrients, such as phosphorus, at different rates. Interference competition occurs mainly among animals, for example, when large males of lake trout chase away smaller males from the best spawning grounds. Competition becomes more important with increasing similarity in resource use, which means that competition is generally stronger between individuals of the same species (**intraspecific competition**) than between individuals of different species (**interspecific competition**). Further, competition is generally stronger between closely related than between distantly related organisms. For example, phytoplankton hardly suffer from competition with fish, whereas competition for nutrients and light with other primary producers, such as submersed macrophytes, may be severe!

As we will see later in this section, organisms that overlap completely in utilization of a limiting resource are not able to coexist according to the **competitive exclusion principle**. However, decreasing the overlap may allow the organisms to coexist if the differentiation is great enough (i.e. the organisms show **resource partitioning**). In many communities it is clear that limiting resources are partitioned among species. For example, two species of *Notonecta* are separated spatially in small ponds: one of the species, *Notonecta undulata*, has the highest densities close to shore, whereas the other, *Notonecta insulata*, is most abundant in deeper areas (Streams 1987; Fig. 4.2). This separation in space may reflect the fundamental niches of the two species. However, interspecific competition may also result in phenotypic changes in resource use or differences in resource partitioning between populations that live alone (**allopatric**) or in

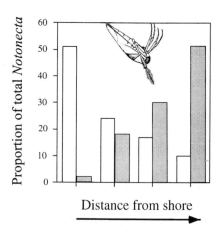

Fig. 4.2   The distribution of two water-boatman species (open bars: *N. undulata*, closed bars: *N. insulata*) in a pond. From Streams (1987).

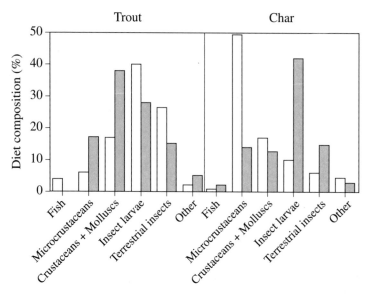

Fig. 4.3   The diet of two salmonid fish, char (*S. alpinus*) and trout (*S. trutta*) when living alone (allopatric, shaded bars) and when coexisting (sympatric, open bars). Both species have similar diets when alone, but char shift to a diet dominated by crustaceans when coexisting with trout. Trout do not change their diet. Modified from Giller (1984); data from Nilsson (1965).

coexistence (**sympatric**) with a competitor. In Scandinavian lakes, allopatric brown trout (*Salmo trutta*) and Arctic charr (*Salvelinus alpinus*) have very similar diets—benthic invertebrates such as amphipods, gastropods, mayflies, and terrestrial invertebrates (Nilsson 1965). When they are sympatric, the Arctic charr shifts to feed mainly on zooplankton, whereas the brown trout continues to feed on benthic prey (Fig. 4.3). A comparison of the habitat use of the two species showed that the dominant brown trout excluded the Arctic char from the littoral zone, at least during the summer (Langeland *et al.* 1991).

Competitive ability

The ability to be an efficient competitor not only differs between species, but also between individuals within a species. Competitive ability is not only a measure of the ability to *reduce* the availability of a limiting resource or to interfere successfully with a competitor, but also includes the ability to *tolerate* reductions in the availability of a contested resource. There are several factors that affect the competitive ability of an organism, including size and excess (luxury) uptake of resources.

Size

Small organisms generally have lower food level thresholds (i.e. can survive at a lower supply rate of a limiting resource than larger organisms). On the other hand, the rate of food collection and assimilation increases faster than the respiration rate with increasing size, suggesting that larger organisms will consume a proportionally larger part of the resources than smaller organisms and will dominate at high resource supply. This can be illustrated with the competition between a large (*Daphnia pulex*) and a small (*Ceriodaphnia reticulata*) zooplankton species in a laboratory experiment (Romanovsky and Feniova 1985). In single-species cultures both species survived at both low and high food supply (Fig. 4.4), but when the two species grew together at low food supply, the smaller species competitively excluded the larger species because the supply rate was not high enough for the large species. At high food supply, however, the large species became dominant.

'Luxury uptake'

Since resource levels fluctuate with time there would be an obvious advantage in assimilating and storing a limiting resource during periods

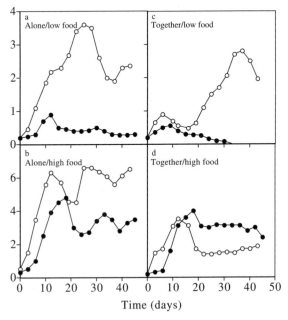

Fig. 4.4 Results from a competition experiment between a large (*D. pulex*, solid symbols) and a small zooplankton species (*C. reticulata*, open symbols). Both species survived in single-species cultures at both high and low food supply (a and b). When grown together, *Daphnia* was excluded at low food supply (c), but became dominant when food supply was high (d). After Romanovsky and Feniova (1985).

when its availability is high and using it when it is scarce. Such 'luxury uptake' is not only common among lower organisms, such as bacteria, fungi, and algae, but also in higher plants and animals. In freshwater systems, the most common limiting nutrient for primary producers is phosphorus. Accordingly, phosphorus is often taken up and stored in specific granules in the form of large polyphosphate molecules. During periods of low phosphorus availability in the environment these macro-molecules are broken down into small phosphate molecules and used in cell metabolism. In this way the organism is able to survive periods when the supply rate of phosphorus would otherwise have been too low for survival. Roots and rhizomes of macrophytes also have a large storage capacity for various nutrients. Storage occurs primarily in vacuoles, isolated from the cell's metabolism. Besides phosphorus, carbon (sugar and starch) is also produced in excess and stored during periods of high photosynthesis. Further, nitrogen may be stored, either as proteins and amino acids or as nitrate. As in humans, many aquatic animals are able to store energy in the form of fat. Fish, for example, store fat before spawning and winter so that enough energy is available for gonad development and spawning behaviour, and for surviving the harsh winter conditions, respectively.

Mathematical models of competition

Man has always, as hunters, fishers, or farmers, been interested in finding answers to questions like: Why is a certain organism increasing or decreas-ing in abundance? Later, the same type of questions became a core issue among ecologists and their mission was to formulate general laws behind fluctuations in abundances of organisms. In the beginning of the nine-teenth century several attempts were made to mathematically formulate such laws. One of the first attempts included the size of the population, the reproductive potential, and the carrying capacity of the environment. Using these three variables will produce a growth curve that initially increases lin-early with the reproductive rate of the organism. Eventually, when resources become depleted; that is, the population is approaching the carrying capa-city (K) of the environment, the growth curve evens out asymptotically. This simple model relatively well describes the increase in abundance of a population of organisms, for example, bacteria in a test tube with nutrient solution. Today, more advanced models are commonly used in ecology, for example, when describing competitive events. The reason for using models is to be able to predict the outcome of competitive interactions. The main importance of models may not be the ability to mimic natural systems, but to shape the way we ask questions and generate ideas, thereby determining the way experiments are performed and data collected. Hence, models should, in addition to experiments and observations, be viewed as tools for improving the understanding of natural ecosystems. One of the more notable models in ecology is the Lotka–Volterra model of competition.

The Lotka–Volterra model

The basis for the Lotka–Volterra model is exponential growth without any limitations, where the change in population size with time (t) equals growth rate × population size. This is expressed in a mathematical form as:

$$dN/dt = rN \qquad (1)$$

where N is the population size of an organism, for example, *Gammarus* sp. (Fig. 3.15), and r is the growth rate. However, since a population will not grow exponentially for very long due to shortage of resources, this model is unrealistic. Therefore, the 'carrying capacity' (K) of the environment, which is the maximum population size the environment can support, was added to the equation:

$$dN/dt = rN[(K - N)/K] \qquad (2)$$

At low population size (i.e. when N is low), the term $[(K - N)/K]$ will be close to unity, and population growth will then be close to exponential as shown by equation (1). Similarly, at very high population size, N will be close to K and population growth will be close to zero.

If a population of another species, for example, *Asellus* sp. (Fig. 3.15), is added to the model, we can explore the outcome of competition for limiting resources between the two crustacean species. However, it is likely that the *per capita* influence of *Gammarus* on *Asellus* is not similar to the influence of *Asellus* on *Gammarus*. An additional variable has to be added to the equation—the competition coefficient (α), which shows the strength of the effect one species has on the other.

If the effect of interspecific competition equals intraspecific competition then the competition coefficients equal unity, and the competition is said to be symmetric. However, this is seldom the case, since competition is usually **asymmetric**. The result is that two different competition coefficients have to be determined: the effect that individuals of *Gammarus* (G) have on individuals of *Asellus* (A; α_{GA}), and vice versa for *Asellus* (α_{AG}). The inclusion of two species leads to one population growth equation for *Gammarus* and one for *Asellus*:

$$dN_G/dt = r_G N_G \left[(K_G - N_G - \alpha_{GA} N_A)/K_G\right]$$
$$dN_A/dt = r_A N_A \left[(K_A - N_A - \alpha_{AG} N_G)/K_A\right]$$

Depending on the values of the variables r, K, N, and α the two species may either coexist or one of the populations becomes extinct and the other eventually increases in size up to the carrying capacity of the environment.

One of the earliest experiments on competition was performed by Gause (1934), using the ciliate *Paramecium caudatum* and growing it in test tubes with bacteria as food. After an initial lag phase (days 0–6; Fig. 4.5), the population of *P. caudatum* grew almost linearly provided that the food (bacteria) was in excess of that needed (days 6–10; Fig. 4.5). Thereafter, growth rate declined owing to lack of food (i.e. the carrying capacity, K, of the test tube was reached). Gause (1934) performed an additional experiment by mixing *P. caudatum* with another *Paramecium* species (*Paramecium aurelia*). Grown in separate test tubes, both species followed similar growth curves (Fig. 4.5). A similar curve was also shown by *P. aurelia* when grown together with *P. caudatum*, indicating that this species was little affected by competition from *P. caudatum*. However, the growth rate of *P. caudatum* was severely reduced when it coexisted with *P. aurelia*, and after 24 days only a few *P. caudatum* remained. This experiment elegantly illustrates the **competitive exclusion principle** which states that if two species occur in the same area and have the same resource requirements, they cannot coexist, one of the species will increase to its carrying capacity and the other will go extinct.

Although the mathematics behind the Lotka–Volterra competition model is relatively simple, the model's practical use is limited, mainly because in natural ecosystems more than two species/populations are generally competing for the same resource. Including more species, by adding more than two equations, will make the model increasingly complicated. Moreover,

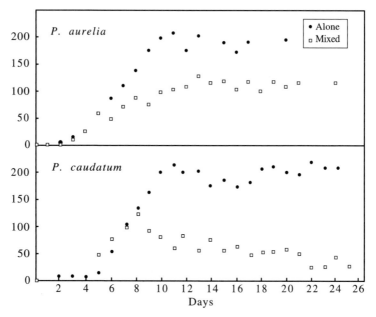

Fig. 4.5   Growth of two species of ciliates cultivated separately and together. *P. aurelia* is only slightly affected when grown together with *P. caudatum*, whereas *P. caudatum* is competitively excluded from the culture in the presence of *P. aurelia*. After Gause (1934).

there are many unrealistic assumptions underlying the model, for example, that all individuals in a population are equivalent (cf. age and size structure in natural populations) and evenly mixed so that they affect each other in a similar way. However, as a first step to model competitive interactions, and for use on laboratory cultures, the Lotka–Volterra model is adequate and stimulating.

The 'two species–two resources model'

One step forward from the Lotka–Volterra model is the 'two species–two resources model' provided by Tilman (1980, 1982), which is a graphical model based on the supply rate of two resources. Briefly, the model can be described as follows. The areas in which species A and B will grow are defined by solid and broken lines, respectively (Fig. 4.6(a)). At the lines, the reproduction rate is balanced by mortality rate and these lines are therefore called the zero net growth isoclines (ZNGI). Hence, if the supply rate of resources x and y for a specific environment falls outside the solid line, the species will become extinct. From the two ZNGI lines we see that species A can suffer lower supply rates of resource y than can species B, whereas the opposite is the case for resource x. If the resource use curves for species A and B are coupled, as shown in Fig. 4.6(b), the two ZNGI lines will cross each other at the two species' equilibrium point (indicated with a black dot). The rate of consumption of both resources by a certain species can be shown as a vector for each species, which is the result of the species' consumption rate of resource x and y (marked A* and B*) in Fig. 4.6(b). This figure can now be divided into six different regions numbered from 1 to 6. Again, in the unshaded area (region 1), the supply rate of resources is

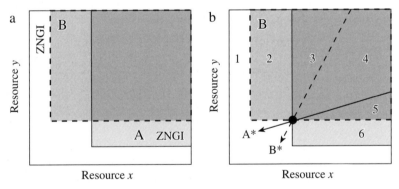

Fig. 4.6 A graphical model of competition between two species for two resources: (a) ZNGI for species A (solid line) and B (broken line). The areas included by the ZNGI indicate the resource supply of x and y within which species A and B, respectively, can grow; (b) ZNGI and consumption vectors identifying the equilibrium point (black dot). The consumption rate vectors and ZNGI's divide the model into six different sectors. A resource supply rate within each of these sectors leads to different outcomes of the competitive situation. See the text for more details. Based on Tilman (1982).

not sufficient for any of the species. In region 2, species B can survive but not species A. In region 3, species B will competitively displace species A, because B will make resource x limiting for A. If the supply rate of the resources in a specific habitat is within region 4, the two species will coexist, because an individual of one species will have a greater effect on its own relatives than on members of the other species. Resource supply rates defined by region 5 will result in species A displacing species B (i.e. opposite to region 3), whereas in region 6 only species A will survive (opposite to region 2). This modelling approach has proven useful under laboratory conditions, and, at present, Tilman's competition model is one of the models that has been most clearly verified in experiments.

'Paradox of the plankton'

According to the concept of competition, the species best able to acquire a limiting resource should displace all other competing species. If this was the case, only one species of, for example, phytoplankton would occur in a lake. Although most species of phytoplankton have similar resource requirements, at least 30 species generally coexist in a lake. This discrepancy between theory and what we can easily see in nature was formulated by Hutchinson (1961) as the '**paradox of the plankton**'. Several sets of explanations for this discrepancy have been put forward, including:

1. *Patchiness of resources.* Resource availability in lake water is not homogeneous. Instead, patches, varying in space and time, favour different plankton species, since some, but not all, species are rapidly able to take advantage of small ephemeral patches. If these patches have a lifetime of less than 10–25 days, which is the time period necessary for competitive exclusion to occur in the phytoplankton community, numerous species will coexist in a non-equilibrium state.

2. *Factors other than competition are important.* A highly competitive species that would have displaced all other species in a laboratory experiment, may be vulnerable to, for example, grazing or sensitive to low pH, temperature fluctuations, or any other variable characteristic for a specific lake. This would make it less competitive, allowing other species to acquire enough resources to coexist with the competitively dominant species.

These two explanations may be enough to accept the conclusion that while complete competitive exclusion may be possible under controlled laboratory conditions, it cannot be expected in natural ecosystems. Instead, it may even be worth turning the 'paradox of the plankton' upside down and instead asking why there are so few species coexisting in a specific lake (Sommer 1989)! Given that there are several thousands of, for example, algal species that could potentially invade a lake, we generally detect less than a hundred! Competitive interactions occur among all groups of organisms and below we will present some examples of such interactions in more detail.

Examples of competitive interactions in lakes and ponds

Periphyton versus phytoplankton: bidirectional resource supply

We have now seen how the ability to acquire resources in the water column is a factor determining which species or group of phytoplankton are dominant. However, algae not only occur in the water column but are also common on surfaces (e.g. at the sediment surface). The resource requirements of these sediment-dwelling (periphytic) algae are not different from those of phytoplankton (i.e. algae in both habitats require nutrients as well as light). In most lakes, the major part of mineralization, the breakdown of dead organisms to nutrients, occurs at the sediment surface. This means that the nutrient availability at the sediment surface is, at least in low and medium productive lakes, higher than in the water. Since light is also an essential resource for algae, the resource supply is bidirectional, light coming from above and the main part of the nutrients from the sediments. Hence, by occupying different habitats, periphyton and phytoplankton have first access to one each of two essential resources. From this simple model, it may be predicted that in lakes of an extremely low productivity, both periphytic and planktonic algae suffer from nutrient limitation. In accordance with these predictions, both algal growth forms show low biomass where productivity is extremely low (Fig. 4.7, area I). When productivity (here expressed as nutrient concentration) increases along the productivity gradient of lakes, the biomass of periphytic algae rapidly reaches its maximum, whereas the phytoplankton biomass is far from at its maximum (Fig. 4.7, area II). This indicates that the nutrient supply at the sediment surface is high and that this supply is consumed primarily by the sediment-dwelling algae. The biomass of phytoplankton continuously increases with increasing productivity, whereas periphyton biomass levels off at moderate productivity, and decreases at high productivity (Fig. 4.7, area III). A likely explanation for this decrease is that the high densities of phytoplankton in the water column reduce the supply rate of light to a level where photosynthesis at the sediment surface is not possible. Without knowing anything about consumption rate of light and nutrients, we may, as a hypothetical experiment, replace resource x and y in Tilman's 'two species–two resources' model (see above) with nutrients and light, respectively, and species A and B with planktonic and periphytic algae, respectively (Fig. 4.8). At a high light but low nutrient supply rate (corresponding to area I in Fig. 4.7), periphytic algae are the dominant primary producers, whereas phytoplankton are close to their ZNGI. In area II in Figs 4.7 and 4.8, the supply of nutrients has increased, whereas the light supply has decreased only slightly since the phytoplankton biomass is not high enough to reduce light penetration through the water. At very high supply rates of nutrients, the supply rate of light has crossed the ZNGI for periphytic algae, which means that periphytic algae will eventually be excluded from the arena (Figs 4.8 and 4.7 area III).

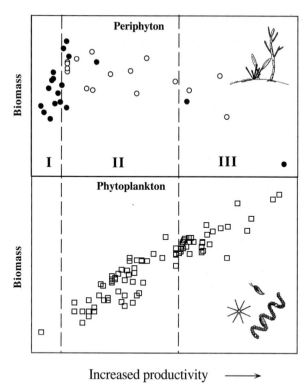

Fig. 4.7 Biomass development (chlorophyll) of periphytic algae at the sediment surface and phytoplankton along a gradient of increasing productivity. At extremely low productivity (area I), both algal life forms increase in biomass when productivity (nutrient availability) increases. Already at moderate productivity (mesotrophic conditions), the biomass of sediment-dwelling periphyton reaches maximum values (area II), whereas phytoplankton continue to increase. The increase continues for phytoplankton at high productivity (eutrophic conditions), whereas the periphyton show a decline in biomass (area III). Periphyton were sampled at 0.75 m water depth. Data are from Antarctic (solid symbols), sub-Arctic, and temperate lakes (open symbols). Data from Hansson (1992*a,b*).

Depth distribution of macrophytes

The bidirectional supply of light and nutrient resources not only affects the biomass development of planktonic and periphytic algae, but may also be involved in determining the distribution of all aquatic primary producers. Submerged and floating-leafed macrophytes absorb nutrients both from the water and the sediment, which means that their supply of nutrients will seldom be growth-limiting. Emergent macrophytes absorb nutrients from the sediment and photosynthesize above the water, a strategy that makes them competitively superior to all other primary producers with respect to acquisition of nutrients and light. Hence, in a situation where competition for nutrients and light is important, emergent macrophytes will win the arms

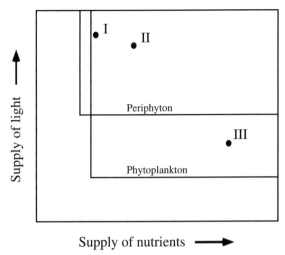

Fig. 4.8 An application of Tilman's 'two species–two resources' model using sediment-dwelling periphytic algae and phytoplankton (Tilman 1982). Because of their position in the nutrient-rich sediment, periphytic algae have prime access to nutrients, whereas phytoplankton have prime access to light. At low supply rate of nutrients (corresponding to area I here, as well as in Fig. 4.7), phytoplankton will be close to their ZNGI. At higher nutrient supply, and only slightly less supply of light to the sediment surface (area II), both life forms will coexist. At very high nutrient supply, the supply rate of light for periphyton will be below the ZNGI (area III) (i.e. death rate becomes higher than growth rate).

race by being unaffected by all other primary producers in the competition for these resources. The biomass development of primary producers along a productivity gradient of lakes can be shown as in Fig. 4.9. Submerged macrophytes and phytoplankton both absorb light in the water column, but the macrophytes take up nutrients from the sediment and, thus will dominate over phytoplankton in less productive lakes. Sediment-dwelling algae are, as we have already seen, the last to get access to light and will therefore have their maximum biomass in shallow areas and in lakes of low productivity. Although emergent macrophytes are superior competitors for nutrients and light, they will generally not become a dominant primary producer in lakes of low productivity, mainly because the amount of nutrients needed by these large plants is not available. However, in medium and highly productive lakes, the nutrient availability in the sediments is high enough to allow for a high relative biomass of emergent macrophytes (Fig. 4.9). Although the ability to absorb nutrients from the sediment and light from above the water surface makes emergent macrophytes competitively superior, it also restricts them to shallow water, making emergent macrophytes a less dominant primary producer in most lakes. The main problem is that the stem cannot be made too tall without the risk of it breaking, or not being able to support the roots and rhizomes with photosynthetic products (sugar) and oxygen.

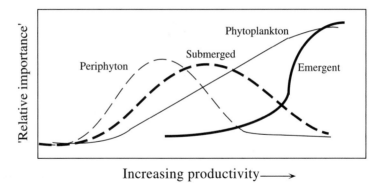

Fig. 4.9   'Relative importance' of different primary producers along a productivity gradient. Sediment-dwelling periphytic algae dominate at very low productivity (see also Fig. 4.7). When productivity increases, submerged macrophytes and phytoplankton increase in importance. At very high productivity only phytoplankton and, in shallow areas, emergent macrophytes, are of importance as primary producers. At extremely high productivity, phytoplankton biomass declines because of self-shading.

As expected, different macrophyte species differ in their ability to grow in deep water (i.e. in their competitive ability), along a depth gradient. This can be illustrated by the segregation between two cat-tail species, the thin, tall-leafed *Typha angustifolia* with large rhizomes, and the more robust *Typha latifolia* with broad leaves but smaller rhizomes. When established at a new site *T. latifolia* is competitively superior in shallow waters, partly due to its broader leaves allowing more photosynthesis (Grace and Wetzel 1981). At a water depth of between 0.6 and 0.8 m, *T. angustifolia* becomes more competitive due to thinner, taller leaves and larger rhizomes with high storage capacity (Fig. 4.10). However, on a longer time perspective *T. angustifolia* generally becomes dominant even at shallow depths (Weisner 1993), mainly

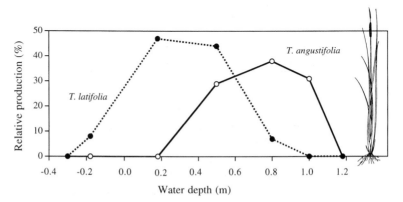

Fig. 4.10   The relative production of two species of emergent macrophytes along a gradient of water depth. *T. latifolia* dominates at shallow depths, whereas at depths deeper than about 0.6 m *T. angustifolia* becomes dominant. Based on data from Grace and Wetzel (1981).

due to a more efficient rhizome support to its taller shoots. Hence, the different traits of these closely related macrophyte species initially lead to a segregation along a depth gradient. In this way, different growth forms of aquatic macrophytes are segregated along depth gradients, which can be seen in most lakes. Closest to the shore, emergent macrophytes dominate, further out floating-leafed and then submerged macrophytes are generally the most abundant (see also Fig. 3.5).

Snails and tadpoles

Snails are an important component of the macroinvertebrate assemblage in many lakes and ponds. Although there may be considerable separation along the spatial, temporal, and food resource axes with different species occupying different habitats and feeding on different food resources, several studies have suggested that intra- and interspecific competition are important structuring forces for freshwater snails. Both laboratory and field studies have shown that an increase in snail density has a negative effect on snail growth and fecundity. Increases in growth rate and fecundity after additions of high-quality food or fertilization to increase periphyton production suggest exploitative competition for food as the regulating factor.

Periphyton is an important food resource for snails which have been shown to control efficiently the biomass of periphyton (Brönmark *et al.* 1992). However, other freshwater herbivores also feed on periphyton and may thus interact with snails for this limiting resource. Chironomid larvae in an Arctic lake were greatly reduced at high densities of the snail *Lymnaea elodes* and this was suggested to be due to not only exploitative competition for periphyton, but also to direct physical interactions between snails and chironomids (Cuker 1983). Grazing snails dislodge chironomid larvae from the substrate and, at least temporarily, disrupt their foraging activity. This is especially serious for tube-building chironomids.

Tadpoles are another important group of periphyton grazers. They occur in all types of water, but Wilbur (1984) argued that in small, temporary ponds desiccation will reduce population densities, whereas in larger, more stable ponds predation will be the most important factor. Only in ponds that predators have not colonized will tadpole populations build-up to such high densities that competition will be important. Numerous studies on different tadpoles in both the laboratory and field have demonstrated effects of intra- and interspecific competition on growth and length of the larval period (Wilbur 1976; Travis 1983), which are correlated with fitness of the adult frogs (Semlitsch *et al.* 1988). Similarly, Lodge *et al.* (1987) suggested that competition between freshwater snails is only important in medium-sized waterbodies. In small, temporary ponds abiotic factors, such as desiccation and winterkill, regulate snail population sizes, and in larger ponds and lakes efficient predators, such as fish and crayfish, will reduce snail populations and be the most important structuring force. Thus, in small ponds without

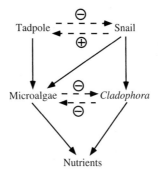

Fig. 4.11   The interaction between herbivorous snails and tadpoles and their algal resources. Solid arrows indicate consumer resource interactions, whereas negative, broken arrows denote competitive interactions and positive, broken arrows facilitation. See text for further explanation. Modified from Brönmark *et al.* (1991).

predators there seems to be a high potential for competition between snails and tadpoles. This was tested in a field experiment in a eutrophic pond (Brönmark *et al.* 1991). Tadpoles had a negative effect on snail growth and fecundity, but snails had a positive effect on tadpole growth and length of the larval period. It was suggested that this facilitative effect was, surprisingly, due to competitive interactions between different types of algae (Fig. 4.11). The filamentous green alga *Cladophora* is competitively superior over periphytic microalgae by reducing the availability of phosphorus and thereby limiting microalgal growth. Tadpoles feed on microalgae, but avoid *Cladophora*, whereas snails feed on both algal groups. Thus, grazing by snails reduced the biomass of *Cladophora* allowing microalgae to thrive which resulted in an increased food resource base and, subsequently, higher growth rate in tadpoles.

Intra- and interspecific competition in fish

Manipulations of densities of a fish species in laboratory and small pond experiments have often shown that intraspecific competition has strong negative effects on growth and fecundity. Crucian carp, for example, is a species that commonly occurs as the only fish species present in the system. It has a unique physiological adaptation that allows it to use anaerobic respiration with glycogen as an energy source during periods of oxygen deficit. Thus, in shallow ponds that experience lack of oxygen during harsh winters (winterkill) it is often the only species that survives. Populations in such ponds typically consist of high densities of small-bodied, short-lived fish, suggesting that intraspecific competition is at work. This is in contrast to ponds and lakes where crucian carp coexist with piscivores; here populations consist of few, large individuals and size-selective predation is the major structuring force in these populations (Brönmark and Miner 1992; Tonn *et al.* 1992). The importance of intraspecific competition in single-species

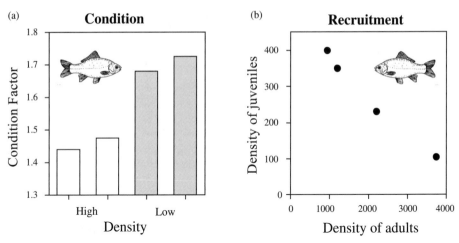

Fig. 4.12   The effects of intraspecific competition in crucian carp. (a) Mean condition factor (100 × mass/total length³) of crucian carp in two high and two low density sections of an experimental pond at the end of the experiment; (b) stock recruitment relation for crucian carp in the experimental pond (i.e. the density of juvenile fish, 0+, versus the density of older fish, >1+). Data from Tonn *et al.* (1994).

ponds was investigated in a field experiment where a pond was divided into sections and crucian carp were added at low and high densities, respectively (Tonn *et al.* 1994). The mortality of large fish was low in all treatments, but at high densities there was a reduced growth rate and the fish were in poorer condition, measured as condition factor (100 × body mass/total length³) (Fig. 4.12). Fish at high densities also had less glycogen stored in their livers and this could lower their overwinter survival rate. Further, there was a strong negative relation between the recruitment of young-of-the-year (0+) crucian carp and the density of older fish (>1+) (Fig. 4.12). This was due to a combined effect of starvation and cannibalism.

However, in most natural systems several fish species coexist. They often have overlapping resource use, at least during parts of their life histories, and this may result in very complex competitive interactions. Most empirical studies of interspecific competition consider pairwise interactions, but recently the complex effects of multispecies interactions have also been considered. For example, Bergman and Greenberg (1994) studied the competitive interactions between ruffe, roach, and perch in eutrophic systems. Ruffe is a specialized benthivore feeding almost exclusively on benthic macroinvertebrates and roach is an efficient forager on zooplankton. Perch, however, undergoes an ontogenetic niche shift with regard to diet, first feeding on zooplankton, then shifting to a diet of benthic macroinvertebrates and, finally, feeding on fish (Fig. 3.22). Efficient foraging for these different kinds of prey requires quite different morphological adaptations and, thus, perch should be a less efficient planktivore than roach and a worse

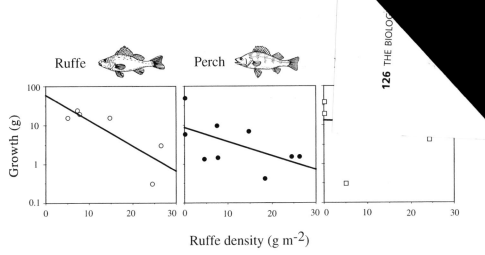

Fig. 4.13 The growth of ruffe, perch, and roach in a gradient of increasing ruffe densities. From Bergman and Greenberg (1994).

benthivore than ruffe. To test the outcome of such three-species interactions, Bergman and Greenberg (1994) studied the growth and diet of these species in an enclosure experiment where they varied the density of ruffe while perch and roach densities were kept constant. Increasing ruffe densities resulted in decreasing growth rates of both ruffe and perch (Fig. 4.13), demonstrating the strong effect of intraspecific (ruffe–ruffe) and interspecific (ruffe–perch) competition for benthic macroinvertebrates. As expected, ruffe had no effect on the zooplanktivorous roach. Other studies have shown that roach forces perch to shift from feeding on zooplankton to macroinvertebrates at a suboptimal size (Persson and Greenberg 1990). Thus, perch is 'sandwiched' between ruffe and roach when these species are present, perch having to compete with the efficient planktivore roach during its planktivorous life stage and with the benthivore ruffe during its benthivorous stage.

Predation and herbivory

In a broad sense, predation can be defined as the consumption of living organisms and thus it includes carnivory and herbivory, as well as parasitism. In terrestrial ecosystems, predation and herbivory usually have different definitions, where predation is strictly defined as when a predator consumes organisms that it has killed. A herbivore, on the other hand, feeds on plants which it rarely kills, but rather consumes only a portion of, without having a lethal effect, such as when a deer eats leaves from a tree. However, herbivores in freshwater systems differ from their terrestrial counterparts in that they generally ingest the whole plant, such as when *Daphnia* feed on algal cells. Hence, with the exception of macrophyte grazers, freshwater herbivores are functionally predators, suggesting that there is less difference between predation and herbivory in freshwater than in terrestrial systems. Therefore, in the following section we will treat predation and herbivory as functionally similar processes with the only difference that predators eat animals, whereas herbivores eat plants. We will often refer to both processes as simply 'consumption'. Below, we will describe in more detail different interactions between consumers and prey common in ponds and lakes, and adaptations that increase foraging efficiencies and prey's ability to survive in the presence of predators. Implications of these interactions for population dynamics and food web interactions in ponds and lakes will be dealt with in Chapter 5, *Food web interactions*.

Principles of predation and herbivory

Predation and herbivory are major sources of mortality in many freshwater organisms and are thus important structuring forces in lakes and ponds, affecting population dynamics and community structure (Kerfoot and Sih 1987; Carpenter and Kitchell 1993). Through direct *lethal effects*, predators may control prey populations and even cause prey species to go locally extinct, and thereby affect prey species distribution patterns and species diversity of a system. Predators may also change the *behaviour* of prey organisms including changes in habitat use, activity patterns, and foraging and this may have indirect effects on prey growth and reproduction rates. Predation can thus be a strong selective force and over evolutionary time many adaptations against predation have evolved in freshwater prey organisms, including, for example, *protective armour*, *toxic chemicals*, and *changes in behaviour*.

The different adaptations in a prey species that decrease its vulnerability to a predator can be termed the *enemy-free space* of the prey (Jeffries and Lawton 1984). Enemy-free space is an important component of a species' niche and many of the patterns in species' distributions that were previously explained by differences in competitive ability can instead be explained by differences in enemy-free space. However, as natural selection improves the prey's ability to escape predation through the evolution of efficient anti-predator adaptations, we could expect a high selective pressure on predators to improve their efficiency, such as increased ability to detect or to capture prey organisms. Thus, over evolutionary time there has been a complex arms race with adaptations and counter-adaptations between predators and their prey. Why then have the adaptations of predators not made them so efficient that they drive their prey to extinction? One suggestion is that prey organisms are ahead in the co-evolutionary arms race due to unequal selection pressures on predator and prey, the so-called *life-dinner principle* (Dawkins and Krebs 1979). This principle was originally described in terms of foxes and rabbits, but we may paraphrase it for freshwater

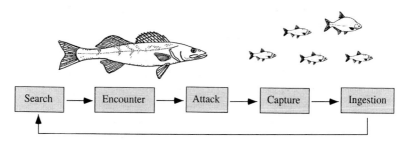

Fig. 4.14   The predation cycle. The different stages of a predation event, from search to ingestion.

purposes as: 'The roach swims faster than the pike, because the roach is swimming for its life while the pike is only swimming for a meal.' A pike that fails to catch a prey may still reproduce, whereas a roach that becomes a meal to a pike will never again return to its spawning grounds. Thus, asymmetries in fitness gains due to the development of a more efficient predation strategy or a more efficient defence, respectively, should result in unequal selection pressures on predator and prey, keeping the prey ahead in the arms race. Further, prey organisms are generally smaller and have shorter generation times than their predators, and can therefore evolve adaptations at a faster rate.

The predation cycle

A predation event, such as when a pike has been successful in capturing a roach, can be characterized as a sequence of different stages—a predation cycle (Fig. 4.14). The different stages include *encounter* and *detection* of the prey, *attack*, *capture*, and, finally, *ingestion*. Efficient defence adaptations have evolved in prey organisms, as have counter-adaptations in predators, at each of the different stages of the predation cycle. Of course, a success-ful predator should have adaptations that allow it to get to the final stage as rapidly and efficiently as possible, whereas a prey should evolve adaptations that interrupt the predation cycle as early as possible. Further, different predators may differ in their efficiency at any one of the stages, and prey species or prey size classes may differ in susceptibility at the various stages.

Feeding modes of herbivores

Herbivores occur both in planktonic and benthic habitats and they are either **filter-**, **surface-**, or **raptorial** feeders. A large group of herbivores are planktonic filter-feeders, such as cladocerans, copepods, rotifers, ciliates, and heterotrophic flagellates, feeding on planktonic algae and bacteria. Benthic grazers include many protozoa (ciliates, amoebae, and heterotrophic flagel-lates), but also larger organisms such as mussels and insect larvae. They live on surfaces, such as macrophyte leaves, on the sediment surface, on stones

Table 4.1 Filtering rates and preferred food particle size (greatest axial linear dimension, GALD) for some important grazers. (Partly based on Reynolds 1984)

	Filtering rate (ml^{-1})	Preferred particle size (μm)
Rotifers	0.02–0.11	0.5–18
Calanoid copepod	2.4–21.6	5–15
Daphnia (small)	1.0–7.6	1–24
Daphnia (large)	31	1–47

and on old car wrecks, and their main food source is the complex microlayer of periphytic algae, bacteria, and detritus on the surface; the *Aufwuchs*.

Filter-feeding

Planktonic filter-feeders include protozoans, rotifers, and crustaceans. As we saw in Chapter 3 (*The organisms*), rotifers have a ciliary corona that creates a water current which manoeuvres particles into the gut. Some rotifers feed primarily on detritus (e.g. *Filinia* and *Conochilus*; Fig. 3.12), whereas others eat small algae and bacteria. However, the major planktonic herbivores are the cladocerans (e.g. *Daphnia*) and copepods, which generally are filter-feeders, although some are raptorial feeders.

Individual filtration rates among planktonic filter-feeders vary more than 1000-fold from 0.02 ml water day^{-1} for small rotifers to over 30 ml water day^{-1} for *Daphnia* (Table 4.1). There are also considerable differences in preferred food particle size. Copepods are far more selective in size preferences than large *Daphnia*, which are able to take advantage of particles ranging from bacteria (1 μm) to large algae (Table 4.1).

Mussels are the most important benthic filter-feeders that move water through the body cavity, removing food particles by using their gills as a filtering apparatus. Mussels may become very abundant in places and may then have a significant effect on phytoplankton (see the *Dreissena* example in Chapter 6, *Biodiversity and environmental threats*).

Surface-feeding

Algae growing on substrates such as macrophyte leaves (epiphytic algae) or the sediment surface (epipelic algae) are exposed to grazing by various surface-feeding organisms, including protozoa, snails, and insect larvae. Most aquatic insects are generalists, eating whatever is available of detritus and algae. They are, however, often functionally separated into **collectors**, which are able to filter-feed, **scrapers** feeding mainly on periphytic algae, and **shredders** able to eat vascular plant tissue (Cummins 1973). All three groups have representatives from common insect taxa such as caddisflies (*Trichoptera*) and mayflies (*Ephemeroptera*). Snails (*Gastropoda*) also feed on surfaces using their rasping tongue, the radula, to remove food particles as they move forward. In dense algal mats, snails may in this way create visible grazing tracks.

Raptorial feeding

Not all algal grazers feed by filtering water or collecting particles. Many cyclopoid copepods, in contrast to their relatives the calanoid copepods (Fig. 3.14), are raptorial feeders attacking individual particles, algae as well as animals. They are often viewed as omnivores or as pure predators. Moreover, some protozoa, such as the rhizopod amoeba *Vampyrella*, feed by attaching to an algal cell and sucking out the contents (Canter 1979), in fact, these amoebae graze on particles larger than themselves!

Grazing on aquatic macrophytes

Terrestrial macrophytes are heavily grazed by a multitude of grazers, but this is not the case with their aquatic counterparts. No fish species in the temperate region use macrophytes as their main dish, although both rudd (*Scardinius erythropthalamus*) and roach (*Rutilus rutilus*, Fig. 3.21) may occasionally feed on submerged macrophytes. Some warm water fish (e.g. some *Tilapia* species) eat macrophytes. This is also the case with the grass carp (*Ctenopharyngodon idella*), a fish species originating from China that is frequently exported to temperate regions to clear ponds from weed. The grass carp efficiently removes macrophytes, but then the ingested nutrients are excreted back to the water, promoting the development of algal blooms.

Emergent macrophytes such as reed (*Phragmites australis*), especially the new green shoots, are subject to substantial grazing from aquatic birds, such as geese, at stopover sites during seasonal migrations. Whether this grazing is of importance for the development of emergent macrophyte beds is still under debate. Submerged and floating-leaved macrophytes are also affected by herbivory from aquatic birds, mainly coots (*Fulica atra*), and swans (*Cygnus* sp.). Moreover, crayfish may, when occurring in high densities, cause catastrophic declines in submersed macrophytes. Although they are highly omnivorous, consuming algae, detritus, and invertebrates, a large portion of their diet is plant material. In addition, crayfish are sloppy feeders and affect macrophytes just by cutting them at the sediment surface. Introduction of crayfish may thus dramatically reduce the biomass of submerged macrophytes in ponds and lakes.

Feeding modes of predators

Two basic foraging strategies are used by aquatic predators. **Ambush** or 'sit-and-wait' predators remain inactive, often within a complex microhabitat, until a prey organism passes by and is detected. Actively searching predators move around in search of a potential prey organism. Of course, a hunter uses a lot of energy by moving around searching for prey, whereas sitting around waiting for prey to appear is energetically cheap. Once a prey organism is detected, however, a sit-and-wait predator may use energy-costly fast-starts and attacks. Generally, ambush predators have a high capture success once they decide to attack, whereas the capture success of actively foraging predators may be rather low.

The foraging mode used is not always a fixed behavioural strategy, but some aquatic predators demonstrate considerable flexibility and adapt their foraging behaviour to differences in, for example, habitat complexity or relative abundance of prey organisms. Ambush predators have a high strike efficiency and a short interaction time with the prey, and it may be advantageous to use this strategy in complex habitats where prey escape more easily. Actively foraging piscivores, such as large-mouth bass and perch have been shown to switch to a sit-and-wait strategy when the complexity of the habitat increases (Savino and Stein 1989; Eklöv and Diehl 1994; Fig. 4.15). Increasing prey densities or prey activity levels increases the encounter rate between predators and prey and this may also cause a change in foraging strategy. In odonates, addition of alternative prey resulted in a switch from active searching to ambush foraging in one species, whereas there was no change in foraging mode in other odonate species (Johansson 1992; Fig. 4.15).

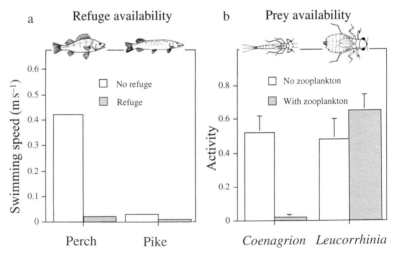

Fig. 4.15   The effect of refuge and prey availability on foraging mode in two predators. (a) Perch decrease their swimming activity in habitats with prey refuges and thus change from an active to a sit-and-wait foraging mode. Pike always use a sit-and-wait strategy, independent of refuges (Eklöv and Diehl 1994); (b) the damselfly *Coenagrion* changes to a sit-and-wait foraging mode in the presence of zooplankton, whereas *Leucorrhinia* practises an actively searching mode, independent of prey availability (Johansson 1992).

Adaptations for prey detection

Aquatic predators may use *visual, mechanical,* and/or *chemical* cues to detect their prey. Predatory fish are mainly dependent on visual cues when foraging, and characteristics of prey organisms, such as size, shape, colour, contrast against the background and movement, may affect the distance at which prey is sighted. Environmental factors that affect light conditions, such as turbidity, may also affect the foraging success of predators that hunt by sight.

Fish may also detect prey organism with their **lateral lines**, a mechanosensory system that is visible externally on fish only as rows of small pores on the trunk and head. The lateral lines allow fish to detect low-frequency water displacements and locate zooplankton during night-time. Experiments have shown that blinded piscivores can locate and attack prey fish at distances of up to 10 cm. A swimming fish generates a complex vortex trail and it has been suggested that other fish can use the information contained in this wake and gain information on, for example, time elapsed since a fish swam by, fish size, swimming speed, and direction. Naturally, this is very useful information for an actively searching piscivore.

Among invertebrates, tactile or mechanical cues are widespread for detecting prey. Vision is generally of less importance in predatory invertebrates, such as larval insects. Aquatic insects have poorly developed eyes and, further, most of them are negatively phototactic (i.e. during daylight they dwell in crevices or under the surface of the substrate). Under these conditions of low light intensity the use of vision to detect prey would be very inefficient and instead the use of tactile cues (using antennae or other mechanoreceptors) is more efficient (Peckarsky 1984). Pond skaters, which forage on the water surface, respond positively to surface vibrations or ripples that are generated by prey organisms trapped in the surface film, but may also use visual cues.

Morphological adaptations for foraging

The morphological design of an organism is, of course, a compromise between different selection pressures, such as selection for efficient foraging, for avoidance of predators, or for reproduction. However, many morphological features of a predator could be directly related to their function in the foraging process. Some of these morphological adaptations act to increase the probability of locating a prey, such as an efficient visual system or the lateral lines in fish, whereas others operate to increase the efficiency of attack or ingestion. Although these different adaptations result in high foraging efficiency on specific prey items, they also result in foraging on other prey items being less efficient or even impossible. Thus, specialized foraging morphologies constrain the predator to a narrow diet.

We often tend to believe that morphological structures are static, but recent studies have shown that in some predators there exists a scope for phenotypic variability in morphological structures associated with foraging. Daphnids grown at low concentrations of food develop larger filter screens than those individuals grown at high food concentrations (Lampert 1994). These filter screens result in higher filtering and feeding rates when both phenotypes feed at identical food levels. This phenotypic plasticity is suggested to be an adaptation to seasonal variations in food availability. Bluegill sunfish from open water habitats in lakes have a more fusiform body shape and shorter pectoral fins than individuals from the more complex, littoral zone with vegetation, where fish have deep bodies and longer pectoral fins (Ehlinger 1990). Littoral morphology is better suited for the slow precise manoeu-vring required for feeding in this zone, whereas open water bluegills are more efficient at cruising. Further, cichlids develop trophic polymorphism of the feeding apparatus or differ in body morphology when they are brought up on different diets (Meyer 1987; Wimberger 1992). In pump-kinseed sunfish, a specialized snail predator, the pharyngeal jaws and bones used to crush snails are more developed in habitats where the major part of the diet consist of snails (Wainwright *et al.* 1991). Thus, in phenotypically plastic species, differential availability or distribution of food resources may result in polymorphism with respect to morphological adaptations for foraging.

Selective feeding

'Selective feeding' means that a predator is feeding on a prey organism more than is expected from its relative availability in the environment. Such selectivity may change a prey community from a dominance of preferred (edible) species to less preferred (inedible) forms, and is a major force behind succession in, for example, algal communities. As seen in Table 4.1, rotifers select smaller sized algae than macrozooplankton. Since the densities of rotifers and macrozooplankton fluctuate over time, we could expect small algae to be less common, compared to large algae, when rotifers dominate over macrozooplankton, and vice versa. This is illustrated by a four-year dataset from Lake Ringsjön, Sweden, which showed a negative relation between the ratio of rotifers ('small') and macrozooplankton ('large') versus the ratio of 'small' and 'large' algae. (Fig. 4.16). Thus, selective foraging by dominant consumers may have far-reaching effects on the entire prey community.

Measurements of selective feeding—stable isotopes and gut content

A traditional way of determining selectivity is to compare what is found in the gut with what is available in the predator's environment. This method is often used and has been useful in many studies. However, one must

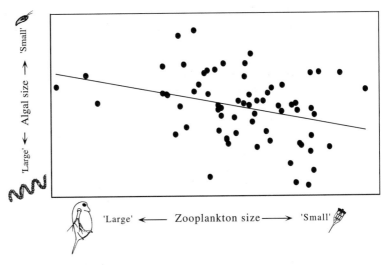

Fig. 4.16 Size relation between zooplankton and phytoplankton communities, showing that a dominance of small zooplankton (high ratio:rotifers/macrozooplankton) is related to high amounts of large algae (low ratio: 'small'/'large' algae). Data from Hansson *et al.* (1998a).

realize that it has several shortcomings. First, there is the problem of prey availability. Naturally, a predator does not encounter prey organisms in the same way as a tube sampler or an Ekman grab, that is, what is really available to a foraging predator is not necessarily the same that we judge available when we sample prey organisms. Further, differences in gut clearance times between prey items, underestimates of soft-bodied prey compared to prey with sclerotized parts, and regurgitation may affect the interpretation of prey selection based on gut contents. Absence of a particular prey item from the diet may not always be an indication of negative predator selectivity, but may also be due to other predator–prey characteristics such as lack of microhabitat overlap or well-developed prey escape mechanisms.

Paradoxically, the absence or low prevalence of a prey item in a predator's diet may actually be due to the predator having a high selectivity for that prey. For example, snails usually make up the major portion of the diet of a specialized molluscivore, the pumpkinseed sunfish, but in lakes with high densities of pumpkinseed fish the snails are kept at such low densities that the fish are forced to include other benthic prey organisms in their diet, and thus the proportion of snails in gut contents is low (Osenberg *et al.* 1992). Similarly, the tench (Fig. 3.21), a common fish in European ponds and lakes, has been suggested to be a generalist benthivore based on gut content analysis. However, when tench were introduced into ponds without fish but high densities of snails and other benthic prey, it turned out to feed almost exclusively on snails until snail densities decreased to lower levels. Hence, in these ponds tench was a specialized forager with a strong preference for snails.

Another drawback with gut content analyses is that the food items found in the gut of a consumer only represents the food it had eaten just before it got caught and thus it only provides a measure of the diet for a short time period and probably from a small part of its total foraging area. To get a more complete picture it is necessary to take samples many times during the growing season and from many microhabitats and this is of course a costly and time-consuming procedure. Further, there are large differences in how well different food items are assimilated. Just because they are found in the gut content does not mean that they are digested and used for growth. Many organisms have adaptations that allow them to pass the alimentary channel of a consumer unharmed. In recent years, analysis of naturally occurring stable isotopes, mainly ^{15}N and ^{13}C, have been used to circumvent the problems with traditional gut content analyses (Rundel *et al.* 1988). In the atmosphere ^{12}C and ^{13}C occur in the ratio of 99 to 1. However, during photosynthesis the primary producers may incorporate ^{13}C in organic molecules to differing extent depending on their physiology but also on the environment where they grow. This will give the plant a characteristic ratio between ^{12}C and ^{13}C called $\partial^{13}C$. The $\partial^{13}C$ change relatively little when the energy from primary producers is moving up the food chain and, thus, the $\partial^{13}C$ of a consumer will reveal the basic source of organic carbon, for example, if it has been fixed in plants within the lake system or if it has been imported to the lake as terrestrial detritus. The ratio of ^{15}N to ^{14}N, called $\partial^{15}N$, on the other hand, increases with trophic level as there is an ^{15}N enrichment of 3–4 per mille with each trophic transfer. Thus, if you know the $\partial^{15}N$ of the basic trophic level, then you may calculate the trophic position of organisms higher up in the food chain. By comparing the stable isotope content of a consumer with its potential prey organisms one could also get a measure of which food items are important for the growth of the consumer over longer time. For example, gut content analysis of crayfish is almost impossible because you cannot identify the well-chewed food items. From observational studies it is, however, known that crayfish feed on detritus, periphyton, macrophytes, and invertebrates, although the proportion of each food item is seldom known. A stable isotope analysis of an experimental food web showed that crayfish is actually a top predator that mainly relies on benthic invertebrates for growth as indicated by a $\partial^{15}N$ value that was 4.9 per mille higher than the invertebrate grazers (Nyström *et al.* 1999, Fig. 4.17). Further, the $\partial^{13}C$ value indicated that primary producers (macrophytes and benthic algae) were the main source of organic carbon in the food web. A drawback with stable isotope analysis is that it does not provide a high taxonomic resolution so a combination with traditional gut content analysis is often very useful. Besides

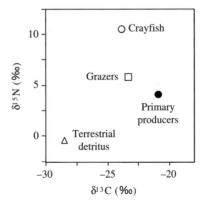

Fig. 4.17   Stable isotope concentrations in terrestrial detritus, primary producers (submerged macrophytes and periphytic algae), invertebrate grazers (crustaceans, insect larvae, and snails), and crayfish. Modified from Nyström *et al.* (1999).

determining consumer diets, stable isotope analysis is increasingly used for determining trophic position of consumers in food web studies and the importance of allochthonous carbon in lake systems (see Chapter 5, *Food web interactions*).

Other approaches to the study of selective foraging in predators are to use predictive theoretical, mechanistic *models*, such as optimal foraging models (e.g. see Stephens and Krebs 1986), combined with manipulative *experiments* and *behavioural observations* of predator–prey interactions.

Optimal diet choice

Optimal foraging theory predicts which prey types a forager should include in its diet if the objective is to achieve maximum net energy gain in order to maximize lifetime reproductive success; **fitness**. The net energy gain of a food particle or prey is the gross energy content of the food minus the energy costs of acquiring it. The profitability of a prey type could be expressed as the energy gained divided by the costs, where the energy costs of handling are often assumed to be proportional to handling time. Thus, profitability is energy gained per unit handling time. The profitability of a prey often increases with prey size because handling time increases more slowly than energy gain with increasing prey size. However, at some critical size, the handling time will increase sharply with further increases in prey size, resulting in decreasing profitability. Thus, for each prey type there will be a specific size at which the profitability is maximized, and given a choice, the optimal forager should preferentially select this prey size. When a forager is exposed to an array of different prey types, classic optimal foraging theory predicts that the optimal forager should prefer to feed on the most profitable prey. When encounter rates with the most profitable prey are reduced due to declining densities, the second most profitable prey should be included in the diet and so on. Thus, the optimal diet is determined by the density of the most profitable prey, whereas increasing densities of low profitability prey should not result in these being included in the diet. The theoretical predictions have been tested many times and often with freshwater organisms: one of the more classic studies on optimal diet selection involves bluegill sunfish feeding on *Daphnia* of different sizes (Werner and Hall 1974).

Patch use

Prey organisms are not evenly distributed in lakes and ponds. Rather, there is a remarkable patchiness in the distribution of prey resources, often coupled to a high heterogeneity and complexity of the habitat as, for example, in the littoral zone. However, even in the seemingly homogeneous water mass of the pelagic zone there is considerable heterogeneity with respect to resource distribution. Zooplankton, for example, are not evenly distributed but often appear in higher densities at specific depths or in dense swarms patchily distributed in the waterbody. A successful predator must be able to locate and utilize patches with high densities of prey and many theoretical and empirical studies have evaluated how aquatic predators use these patches. Theoretical studies have suggested a number of possible behavioural 'rules' that may allow an individual predator to maximize the time it spends in patches with high densities of prey, and to minimize the time it spends in low prey density patches. Predators that forage in an efficient way aggregate in high quality patches (aggregative response) and to achieve this they may use different, more or less complicated, behavioural strategies. *Area restricted search* is a simple strategy that results in an efficient use of high-quality patches, and it involves an immediate adjustment of the search path in response to changes in prey densities. Thus, a predator that enters a high quality patch will change its movement pattern by increasing the rate of turning or decrease the rate of movement. Figure 4.18 shows a typical example of area restricted search. In this case it is a herbivore, the snail *Lymnaea stagnalis*, that changes its movement pattern in response to changing levels of periphyton biomass, but predators could behave similarly.

Other behavioural strategies involve individual decisions about how long to stay in the patch, and when to give up and move on to search for another patch. A predator may, for example, use a *constant giving up time* (i.e. it abandons a patch a set time after the last prey was encountered). The *marginal value theorem*, on the other hand, predicts that an optimally foraging predator should forage in a patch just long enough to reduce the resource level of the patch to the average value of the habitat as a whole.

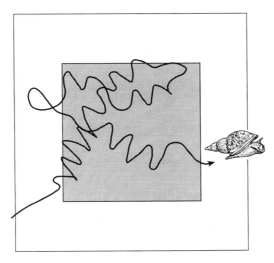

Fig. 4.18   Area restricted search in the snail *L. stagnalis*. When the snail enters a patch with high algal biomass (shaded area) it decreases its rate of movement and increases its rate of turning.

Trade-offs—foraging or avoiding predators?

Although finding and selecting food items in an efficient way are of utmost importance for an animal, other factors are also crucial in determining an individual's fitness. For example, efficient ways of avoiding predators and finding a partner for reproduction also increase fitness. Animals are less vigilant and more conspicuous to a predator when they move around in search of food. Decreasing activity or changing to a habitat with fewer predators may, of course, reduce the risk of predation, but this often comes with a cost in terms of lost opportunities to feed. Reduced activity levels and/or changes in habitat choice in response to presence of predators have been shown to reduce prey foraging and growth rates in, for example, crayfish, bluegill sunfish, and notonectid backswimmers (Stein and Magnuson 1976; Sih 1980; Werner *et al.* 1983). The relative benefits of avoiding predators and foraging are of course state-dependent: they change with prey vulnerability as determined by, for example, relative size and predator density, food abundance, competition, and hunger levels in the prey animal. Several studies have shown that animals are capable of judging the relative costs and benefits of foraging and avoiding predators under different circumstances and to make a compromise—a *trade-off*—that balances the foraging returns against risk of predation. Milinski (1985) showed that sticklebacks feeding on tubificid worms changed their feeding behaviour in response to predation threat. In the presence of a predator (a piscivorous cichlid) they started to forage at a slower rate and avoided worms that were close to the predator (Fig. 4.19). Notonectid backswimmers are able to adjust their foraging behaviour in a risky environment in response to size and density

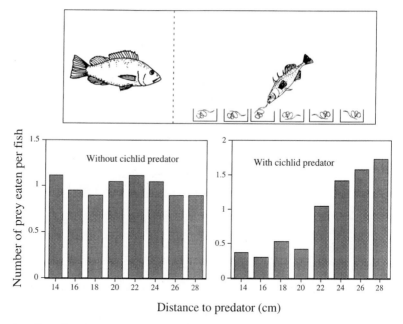

Fig. 4.19   The effect of predator presence on the foraging rate of a three-spined stickleback. The graphs shows the mean number of *Tubifex* taken at different distances from the predator enclosure in tanks with or without a predatory cichlid. From Milinski (1985).

of cannibalistic conspecifics (e.g. Sih 1980). An increased need for food (i.e. increased hunger level), should also affect the optimal compromise between feeding and avoiding predators. Hungry animals should be more willing to take risks, whereas satiated individuals can afford to remain inactive in refugia. Experimental studies have shown that freshwater organisms indeed take bigger risks as hunger levels increase (e.g. see Milinski and Heller 1978).

Functional response of consumers

The number of prey eaten by an individual predator will increase with increasing prey densities, but at some prey density the increase in predation rate will decelerate because of time constraints imposed by handling and ingestion of individual prey. This relationship between the predation rate of an individual predator and prey density is called the **functional response** of the predator. The shapes of the response curves differ between consumers and can be classified into three different categories (Holling 1959; Fig. 4.20). The **type I** functional response represents predation rates of foragers that increase their prey consumption rate linearly up to a maximum level, where it remains irrespective of further increase in prey

FUNCTIONAL RESPONSE

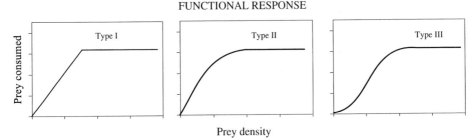

Fig. 4.20   Functional response of consumers (i.e. the relationship between prey density and the per capita predation rate of the predator). Three idealized response curves are shown: Holling's types I, II, and III. See text for further discussion.

density. Type I functional response curves are typical of, for example, filter-feeders, such as daphnids or mussels, where the feeding apparatus sets an upper limit on how many prey can be handled per unit time (Fig. 4.21). In **type II** functional responses the predation rate increases at a decelerating rate until an asymptote is reached where consumption rate remains constant, independent of further increases in prey density. The exact shape of the functional response and the level of the plateau depend on characteristics of both the predator and the prey. The predation rate of backswimmers (*Notonecta*) in response to different characteristics of prey organisms has been studied in great detail (e.g. see Fox and Murdoch 1978; Murdoch *et al.* 1984). *Notonecta* fed at a higher rate when feeding on smaller prey than on larger, and large individuals had a higher predation rate than smaller ones. This difference in feeding rates is due to larger predators having larger guts but also because larger notonectids were more active and searched

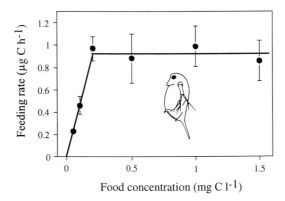

Fig. 4.21   Type I functional response of the filter-feeding *Daphnia pulicaria*. Modified from Lampert (1994).

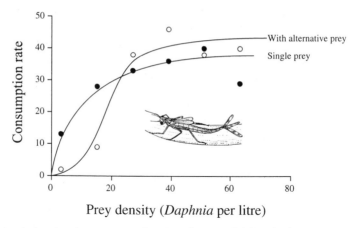

Fig. 4.22   Change in functional response as a function of prey availability. The functional response of a damselfly changed from type II to type III when presented with alternative prey. Curves are fitted by eye. Modified from Acre and Johnson (1979).

a larger area for prey. Well-fed individuals (at high prey densities) had a faster growth rate, resulting in increasing attack rates. Such changes in a predator's attack rate due to growth and development is known as the **developmental response**.

The **type III** functional response is typically associated with vertebrate predators, which have a learning component in their foraging behaviour. Type III functional response curves have a sigmoidal shape with an accelerating phase at lower prey densities and then a decelerating phase which results in a maximum consumption level as in type II functional responses. Generally, a type III response curve will occur when an increase in food density results in increasing search efficiency of the consumers or a decrease of their handling time. A type III response curve may also be generated when the predator is capable of switching between alternative prey items. This can be exemplified with damselfly naiads (*Anomalagrion hastatum*), which had a type II response when only one species of zooplankton prey were present, whereas when they had two cladoceran zooplankton (*Daphnia* and *Simocephalus*) to choose between, the response curve was of type III (Fig. 4.22).

Consumption rate of small organisms

Because of factors such as *selective feeding, consumer-resistant morphologies*, and the composition of organisms in the *prey community*, consumption rates differ considerably both spatially and temporally. The rate at which a specific prey is consumed can be determined experimentally by removing the consumer (herbivore or predator) from a volume of water, then using small enclosures in which known numbers of consumers can be added. A gradient from, for example, 0.5 to 10 times the ambient abundance of the consumer may be appropriate. This type of experiment has been used in particular for zooplankton feeding on algae, but can also be applied to other consumer–prey interactions. The net growth rate (r) can be calculated as the ratio between the final (end of experiment) and the initial (start of the experiment) abundance of the prey divided by the time of the experiment. Mathematically, it looks like this:

$$r = \ln(N_t / N_0)/\Delta t$$

where N_0 is the initial abundance of the prey, N_t is the abundance at the end of the experiment, and Δt is the duration of the experiment (Lehman and Sandgren 1985). The net growth rate (r) for a specific prey species in each enclosure may then be plotted against the predator biomass. If the slope of the linear relation between consumer biomass and net growth rate of prey is steep, the consumption rate is high (Fig. 4.23), whereas a line parallel to the x-axis indicates a negligible impact of the consumer on that specific prey organism. Hence, the slope is a direct estimate of the species-specific consumption rate. This method is very useful in determining consumption rates for most small organisms, including zooplankton grazing on phytoplankton, heterotrophic flagellates grazing on bacteria, or predation of copepods on rotifers. For larger organisms the scaling of the experiment may become a problem.

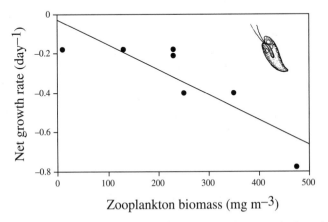

Fig. 4.23   The net growth rate of *Cryptomonas* spp. populations along a gradient of zooplankton biomass. The slope of the relation between net growth rate (r) and zooplankton biomass is an estimate of the filtering ('grazing') rate of the zooplankton on *Cryptomonas* spp.

Defence

Adaptations that reduce the risk of being the victim of predation can be divided into two major categories based on whether they act before or after a predator–prey encounter (Edmunds 1974). **Primary defences** operate before a predator has detected the prey and decrease the probability of an encounter with a potential predator. **Secondary defences** come into action after the prey has been encountered by a predator. The function of the secondary defence is to increase the probability of a prey surviving after being detected by a predator. The different defence adaptations are not mutually exclusive; one and the same prey organism can have several defence adaptations that operate at different stages of the foraging cycle.

Primary defences

Avoiding encounters

Many prey organisms are poor at escaping predator attacks and have a high probability of being attacked and eaten once encountered. Obvious examples are most algal species, as well as bacteria. If we focus on animals, the presence of predators may result in very low population densities or even extinction in prey with poor defence or escape adaptations. Thus, these species need to reduce predator encounter rates. The most efficient way to reduce encounter rates is, of course, to completely avoid systems with predators. Ephemeral ponds, for example, are not inhabited by fish and by spending their larval period here, prey organisms will substantially reduce the risk of predation, albeit with a cost—the risk of not completing the larval stage before the pond dries out (see also Chapter 2).

Prey organisms living in permanent waterbodies may also avoid systems with predators. The tree frog (*Hyla arborea*) is a threatened species in large parts of western Europe, mainly due to changes in the landscape, such as wetland draining and pond-filling. A study of populations of tree frogs in southern Sweden revealed that this frog did not reproduce successfully in all ponds that were available within its distribution range (Brönmark and Edenhamn 1994). A pond where frogs had a high breeding success could be located just next to another pond that was apparently similar but had no calling males in the spring. A survey revealed that all ponds with successful frog reproduction had either no fish, or only sticklebacks, whereas those ponds without breeding tree frogs were consistently inhabited by fish (Fig. 4.24). Thus, tree frogs do not use ponds with fish for reproduction. The exact mechanism underlying this is not known but frogs generally return to breed in the pond where they originated and, thus, a high predation pressure in a pond results in no or few metamorphosing froglets and no adults that can return to the ponds during reproduction. Alternatively, adult females may avoid laying their eggs in ponds with predatory fish, for

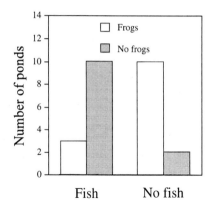

Fig. 4.24   The distribution of reproducing tree frogs (*H. arborea*) in ponds with and without fish. Open bars represent ponds with reproducing tree frogs, and shaded bars are ponds without tree frogs. From Brönmark and Edenhamn (1994).

example, by breeding in temporary ponds only or by being able to detect the presence of fish in the pond.

Many prey organisms are able to coexist with predators although they have poor escape or defence adaptations. In order to do so they must have some mechanisms that minimize the frequency of encounters with predators within the system. Selecting a habitat where predator density/predation efficiency is low (**spatial refuge**), being inactive when predators are active (**temporal refuge**), or having a colour that blends in with the background (**crypsis**) are examples of different ways of reducing predator–prey encounter rates. Below, we will go into more detail regarding these different strategies.

Spatial refugia

A prey organism can minimize mortality from predation by occupying a habitat within a lake that is avoided by predators, for example, due to low oxygen concentrations. However, a habitat where predators coexist with prey may still provide a spatial refuge from predation. This is typical for habitats with high **structural complexity**, where the efficiency of the predator is reduced. A large number of theoretical and experimental studies in both terrestrial and aquatic environments have shown that increasing habitat heterogeneity reduces the strength of predator–prey interactions and allows coexistence of predator and prey by providing refugia for prey organisms. In ponds and lakes, the littoral zone is generally the habitat with the highest structural complexity. Along exposed, windswept shores this complexity derives from coarse, stony substrates. On less exposed sites, where sediments accumulate, submerged and emergent macrophytes provide habitat complexity.

Descriptive studies of the distribution of benthic macroinvertebrates have shown a greater abundance and species diversity of macroinvertebrates in macrophyte beds (Soszka 1975) compared to more barren habitats. Macrophyte beds consisting of different macrophyte species with different plant architecture, and a combination of submerged and emergent macrophytes should provide a higher complexity than monospecific stands of, for example, cat-tails.

A number of studies have manipulated habitat complexity, such as macrophyte density, in enclosures placed in ponds and lakes, while keeping the predation pressure constant. The results from these studies clearly show higher densities and more species of benthic macroinvertebrates in treatments with a high habitat complexity (Gilinsky 1984; Diehl 1992). An increase in habitat complexity results in decreased encounter rates due to lower visibility of prey but also because the swimming speed of the predator decreases, which leads to an increased search time. High habitat complexity also increases the probability that the prey can escape once encountered and may also affect the handling efficiency of captured prey.

Temporal refugia

To reduce encounter rates, a prey may choose to decrease its activity when predators are present to reduce its conspicuousness. Moving prey has a higher probability of being detected and captured by predators. Tadpoles detect the presence of predators by chemical cues and become less active (Petranka *et al.* 1987; Laurila *et al.* 1997) and crayfish reduce their activity and increase their use of shelters when under threat of predation by fish (Stein and Magnuson 1976). A prey may also change its activity pattern so that it is active during a time of day or part of the season when the predator is inactive or absent from the habitat. **Diel vertical migration** and **diapause** are examples of temporal escapes by prey organisms over a short and long scale, respectively.

Diel migration   If you go to a lake and take zooplankton samples at different depths over a 24-hour period you will probably find that cladoceran zooplankton such as *Daphnia* are concentrated at different depths during day and night (Fig. 4.25). This is because many freshwater zooplankton, especially cladocerans, migrate vertically throughout the water column on a daily basis, moving downwards at dawn into the darker bottom layers and then moving up into the epilimnion again at dusk. This large-scale, migratory pattern has intrigued freshwater biologists for many years and many hypotheses have been put forward to explain the adaptive value of this behaviour. It became clear relatively early that the proximal cue triggering this migratory behaviour is change in light intensity, but the ultimate cause was less obvious. It has been suggested to be either an adaptation for behavioural thermoregulation where a downward move into cooler water would

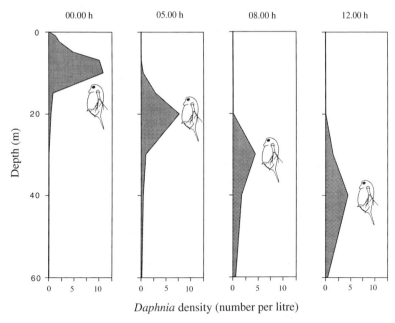

Fig. 4.25   Diel vertical migration of *Daphnia* in Lake Constance in July 1977. From Stich and Lampert (1981).

optimize metabolism, or a means of exploiting high-quality food resources, phytoplankton food quality being highest during night. However, comparisons of migrating and non-migrating populations of daphnids have shown that migration does not confer a metabolic or demographic advantage; *Daphnia* had a more rapid growth rate and produced more offspring when they did not migrate (Dawidowicz and Loose 1992). Rather, spending energy swimming down from the benign environment of surface waters with high temperature and richness of high-quality food down to the cold bottom waters where food resources are scarce and of low quality (mostly detritus) seems to be a maladaptive behaviour hard to explain. The adaptive value of downward migration during the day has been suggested to be a reduction in the mortality that is imposed by visually foraging predators (Stich and Lampert 1981), that is, the dark bottom waters provide a refuge from predation during the day when predation rates by visual predators are at their highest. The fitness costs from spending the days in cold temperatures and poor food conditions, could be compensated by benefits of decreased mortality due to predation.

The strongest evidence for the hypothesis that diel vertical migration is an adaptation to avoid predation comes from a number of very carefully designed laboratory and field studies where vertical migration in cladoceran zooplankton was induced by the presence of chemical predator cues (see

Lampert 1993 for a review). Water from aquaria with fish has been shown to contain a fish **kairomone** that elicits diel vertical migration. A kairomone is a substance released by one organism, in this case the fish, favouring another organism, in this case the zooplankton. The exact identity of the kairomone is still unknown. Different fish species, including zooplanktivorous, benthivorous, and piscivorous fish, release the kairomone, independent of what they have been feeding on. A rapid release of kairomones by fish in combination with their rapid microbial degradation results in a predator cue that is highly reliable as an indicator of fish presence.

It is not only cladoceran zooplankton that show a change in migratory pattern when exposed to chemical cues from their predator. The phantom midge, *Chaoborus* sp. (Fig. 3.20), moves down to deeper waters and remains in the bottom sediment during the day when exposed to fish kairomones (Dawidowicz *et al.* 1990). Interestingly, Neill (1990) found a reversed vertical migration pattern in a calanoid copepod, which moved upwards in the water column during the day and then moved down to deeper waters at night. This behaviour was triggered by the presence of the invertebrate predator *Chaoborus* and Neill was able to show that the cue was chemical; filtered *Chaoborus* water elicited the same migratory patterns as individual *Chaoborus*. *Chaoborus*, in turn, showed a normal diel vertical migration and, thus, the reversed migratory pattern in the copepod could be explained as an adaptive response to decrease predation rates by *Chaoborus*.

Zooplankton in shallow, non-stratified lakes do not have the option of moving down to the metalimnion and hypolimnion to avoid predation. However, studies in these lakes have shown that zooplankton aggregate in nearshore areas among the structurally complex macrophyte beds during the day. For example, the density of *Daphnia magna* was 20-fold higher within the macrophyte beds during daytime than during night-time (Lauridsen and Buenk 1996), indicating that horizontal migration between the structurally complex macrophyte beds and the open water may be a way to reduce the risk of predation by planktivores in lakes where vertical migration is restricted. In parallel to vertical migration, zooplankton exposed to chemical cues from fish increase their use of the macrophyte habitat (Lauridsen and Lodge 1996).

Behavioural responses to avoid being eaten are generally associated with animals. However, some flagellated algal species (e.g. *Gonyostomum* and *Peridinium*, Fig. 3.6) stay at the sediment surface if the density of zooplankton in the water is high (Hansson 2000). When the risk of being eaten is lower, they rise in the water column in large numbers. In a laboratory study the alga *Gonyostomum* entered the water column in high numbers in control treatments without zooplankton (Fig. 4.26), whereas only low amounts of the algae left the sediment in the presence of caged, live zooplankton. Similar results were obtained when the zooplankton were dead, excluding

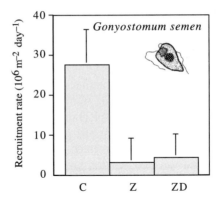

Fig. 4.26 Recruitment of the algal species *Gonyostomum semen* from the sediment to the water column in a laboratory experiment. In controls (C), where no zooplankton were present, many *G. semen* left the sediment and entered the water column, whereas in the presence of living (Z) or dead *D. magna*, the recruitment rates were low. Data from Hansson (1996).

the possibility of direct grazing on the algae. The likely cause for this behaviour is that zooplankton excrete a chemical substance into the water which causes the alga to change its behaviour accordingly.

Diapause   So far, we have considered temporal predator refuges that are rather small-scale (i.e. changes in behaviour over a daily period). However, prey with predators that are periodically absent from the system or that have periods of low activity, may use **diapause** as an adaptation to decrease predation (see Gyllström and Hansson 2004 for a review). Diapause is a transient interruption in growth or development. Hairston (1987) showed that a pond-living copepod responded to changes in sunfish predation pressure by producing different kinds of eggs. In early spring, the copepod produced normal eggs that hatched immediately, but as the activity of the sunfish increased with increasing water temperature, the female copepods began to produce diapausing eggs instead. These eggs did not hatch until autumn when the water temperature dropped and fish activity declined. A natural experiment further strengthened the adaptive explanation of diapause as an anti-predator adaptation. A catastrophic drought killed all the fish and within two copepod generations the timing of the diapause had shifted to one month later in the season allowing the copepods to exploit the high resource levels and high temperatures during summer.

Crypsis   Most freshwater organisms have a dull, colourless appearance, especially when compared to marine organisms on, for example, tropical coral reefs. Some freshwater fish may be brightly coloured, especially during the breeding season, but perhaps the only brightly coloured invertebrate is the red water mite. It has been suggested that this red colouration is a warning signal that advertises their unpalatability due to chemical toxins (Kerfoot 1987). However, few freshwater organisms stand out against their background and many are even camouflaged and blend in with the environment; they are **cryptic**.

Being three-dimensional, animals are often rounded in cross-section and with light normally coming from above, there is a ventral shadow on the body which could be conspicuous to a predator. Many cryptic animals are counter-shaded (darker dorsally, paler ventrally) so that this ventral shadow is obscured. Further, pigmented organisms are more easily located by a visual predator. Zooplankton are typically transparent in lakes with fish, whereas in environments without fish, such as in Arctic or Antarctic lakes and ponds, zooplankton may be quite heavily pigmented (Luecke and O'Brien 1981). Female zooplankton carrying eggs are more visible to fish and suffer more from predation than individuals without eggs (Hairston 1987). A large, pigmented eye also increases the probability of being detected by a predator and it has been shown that zooplankton that coexist with fish have smaller eyes than conspecifics from systems without fish (Zaret 1972).

Secondary defences

If a prey is detected by a predator and the predator attacks, prey organisms may still have several traits that increase the probability of a successful escape. Many prey are capable of rapid escapes or flight which are enhanced by, for example, staying close to a refuge or by having *erratic escape movements*, making it impossible for the predator to predict the direction of the next move. Whirligig beetles, for example, usually swim around on the water surface in loose schools, but when disturbed by a predator they start to swim in a rapid series of erratic movements. The rotifer *Polyarthra* also makes sudden 'jumps' when attacked by a predator, such as a copepod. Other prey organisms have a *retreat* that they can disappear into when approached by a predator. Crayfish dig out burrows in the sediment or hide in the interstices of rocky habitats. Snails and mussels move into their shells and remain there protected by the hard shell. Caddisflies construct cases that they move around in or make tubes and retreats that are attached to solid substrates. The molluscan shells and the caddis cases may also be seen as morphological defence adaptations that decrease the probability of ingestion once they have been captured by making them more difficult to handle.

Spines and other morphological structures may also increase predator handling times, resulting in the predator rejecting the prey or the prey being avoided. Sticklebacks have spines that can be locked in an upright position, creating difficulties for piscivores when handling them. In a classic experiment, Hoogland *et al.* (1957) showed that when given a choice between minnows without spines, 10-spined sticklebacks with small spines and 3-spined sticklebacks with large spines, pike preferred to feed on

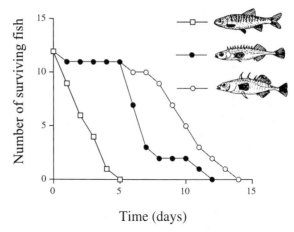

Fig. 4.27 Predation rates of pike (*Esox lucius*) feeding on minnows (*Phoxinus phoxinus*), 10-spined sticklebacks (*Pygosteus pungitius*) and 3-spined sticklebacks (*Gasterosteus aculeatus*). Modified from Hoogland *et al.* (1957).

minnows, whereas 3-spined sticklebacks survived the longest (Fig. 4.27). Absence of spiny fin rays in combination with a more slender body morphology also explained the northern pike's preference for fathead minnows over bluegill sunfish (Wahl and Stein 1988).

Many of the larger, active organisms of the littoral zone use a *chemical defence* against predation by fish (Scrimshaw and Kerfoot 1987). Dytiscid and whirligig beetles, backswimmers and water-boatmen (Fig. 3.19) have glands producing secretions that are discharged when attacked by fish. Numerous chemical compounds have been identified in such secretions, including steroids, aldehydes, esters, and organic acids. Many of these secretions are extremely distasteful to fish and are a very efficient defence. For example, dytiscids that are captured by fish release their secretion, resulting in immediate rejection and vomiting by the fish. Other compounds released by a captured prey have an anaesthetic effect on fish. However, these chemical compounds may also have a negative effect on the prey organism, since they may be quite costly to produce. Interestingly, a recent study has shown that predation rates by diving bugs and damselfly is reduced by about 50% when the prey, an amphipod, has been feeding on a plant that is known to be rich in phenolic compounds (Rowell and Blinn 2003). Use of secondary plant chemicals as a predation defence is well known in terrestrial herbivores but this is the first example from freshwater systems.

Consumer resistance among smaller organisms

Smaller organisms, such as algae and bacteria, may rarely be able to avoid encounter or take advantage of refuges or crypsis. However, it is in the interest of small prey organisms to be as difficult as possible to catch and handle or to taste as unpleasant as possible in order to reduce the risk of being eaten. Herbivores may reject food particles due to several reasons, including *toxicity*, unmanageable *size* or *shape*, and in an evolutionary perspective, prey organisms should take advantage of these possibilities. The most vulnerable particle size for herbivorous zooplankton is a 3–20 μm sphere without spines or any other protection. Some algae, such as *Cryptomonas* and *Cyclotella*, are close to this description and may be categorized as highly edible. The maximum particle size a filter-feeder can ingest is linearly correlated with animal body (carapace) size (Fig. 4.28), indicating that the larger the alga is, the fewer the herbivores that can eat it. Hence, one way to reduce the grazing pressure is to grow *large*, thereby being vulnerable only to the largest grazers. On the same theme is the formation of *colonies*, that is, several small cells attach to each other, and, 'voilà' a food package is created that can only be eaten by the largest of grazers! A striking example of this behaviour is the formation of bundles by the cyanobacteria *Aphanizomenon* (Fig. 3.7). During blooms, these bundles are so large that they can be seen by the naked eye; they look like cut grass spread over the lake (Andersson and Cronberg 1984). Grazing-resistant morphologies are also found among bacteria, which are able to form large spirals and filaments (Jürgens and Güde 1994). The opposite strategy is to become *too small* for the grazer. The smallest single-celled bacteria are not grazed by macrozooplankton and will therefore dominate over larger rod-shaped and filamentous bacteria in situations of high grazing pressure. Just as was the case with sticklebacks (see above), many algal species form *spines*, making the cell difficult to handle for a filter-feeder.

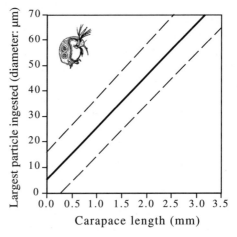

Fig. 4.28   The relationship between body size (carapace length) and the diameter of the largest particle ingested by seven species of Cladocera. Broken lines are 95% confidence limits. Based on Burns (1968).

Some algae (e.g. *Microcystis, Aphanizomenon,* and *Gymnodinium,* Figs 3.6 and 3.7) have been shown to be *toxic* to grazers. However, the quantitative importance of this adaptation is not well known.

Some algae are covered with a *mucilaginous sheet* protecting them from gut enzymes, allowing algae to pass through the gut of the grazer undamaged. Indeed, some of these algae are even able to absorb nutrients during their passage through the gut. It goes without saying that such an adaptation allows algae to benefit from grazing, both due to the loss of competitors digested by the grazer and to the access to a unique nutrient source! This adaptation was first demonstrated by Porter (1973) in enclosure experiments where she removed grazers larger than 125 μm (mainly *Daphnia* and copepods) in some of the enclosures, whereas grazers were added to other enclosures. The main result was that populations of small algae were suppressed in the presence of grazers. These algae have no adaptation to protect them from grazers and may be viewed as highly edible. Surprisingly, some small colonial green algae (e.g. *Sphaerocystis*), were not reduced in the presence of grazers, and in some cases even increased in abundance. Although ingested by the grazer, they were not assimilated due to their protective mucous sheet and emerged growing healthily on faecal pellets!

It should be noted that absolute grazing resistance is rarely achieved, but even a slightly reduced vulnerability may have a profound impact on the grazing pressure that a specific organism suffers from.

Chemical warfare—the cyanobacteria versus Daphnia arms race

The eutrophication process (see Chapter 6) has lead to a worldwide increase in the frequency and intensity of phytoplankton blooms ('green lakes'), and especially to an increase in abundances of cyanobacteria. The most common bloom-forming cyanobacterial species, such as *Microcystis* and *Anabaena* (see Fig. 3.6), form large colonies and are therefore difficult for zooplankton to ingest. In addition to their large size, these algae may excrete extremely potent toxins, poisonous not only to grazing zooplankton, but also to dogs, cows, and of course, humans drinking or swimming in the water. Zooplankton exposed to toxic cyanobacteria either show reduced growth and reproduction or, depending on the concentration of toxins, die (Lürling and van Donk 1997; Gustafsson and Hansson 2004). Of course, such a chemical warfare is beneficial to the cyanobacteria, which can continue growing without much interference from their grazer. However, at least some of the grazers, such as *Daphnia* (see Fig. 3.13), seem to be able to strike back by, at least partly, getting used to the toxin and continue feeding (Gustafsson and Hansson 2004). That this arms race is important also evolutionary was elegantly shown in a study where resting eggs (ephippia; see Fig. 3.13) from *Daphnia* were isolated from different sediment depths; that is, different time periods (Hairston *et al.* 1999). The ephippia were then hatched in the laboratory and exposed to cyanobacterial toxins. The study showed that hatched *Daphnia* originating from the time periods when cyanobacterial blooms were common were more resistant and grew better than *Daphnia* originating from periods when algal blooms were rare (Hairston *et al.* 1999). Hence, if daphnids are pre-exposed to the toxin, they are able to counteract the chemical warfare reasonably efficient in this example of evolutionary arms race.

Constitutive or inducible defences?

Sticklebacks always have their spines, snails their shells, and crayfish their strong chelae, independent of where they live and who they live with, that is, these defence adaptations are present irrespective of whether the predator is there or not: they are **constitutive defences**, always present. However, recent studies have shown that several freshwater organisms have morphological defence adaptations that are phenotypically induced by the presence of a predator (Fig. 4.29). **Inducible defences** only evolve under quite special circumstances and are dependent on certain characteristics of both the predator and the prey (Adler and Harvell 1990). Thus, the evolution of inducible defences is promoted when:

1. There is a *variability* in predation pressure (i.e. the predator should not always be present in time or space). If the predator is always present in the habitat it would of course be more advantageous to have a constitutive defence.

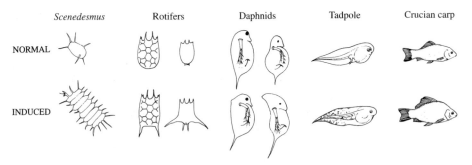

Fig. 4.29 Examples of organisms that have inducible defences (i.e. they show a phenotypic change in body morphology in response to the presence of a predator). The upper row shows the morphology of the organisms in the absence of predators and the lower row in the presence of predators. Cells of the alga *Scenedesmus* join to form large, multicell units in the presence of herbivores (Hessen and van Donk 1993), rotifers form spines in the presence of other, predatory rotifers (Stemberger and Gilbert 1987), daphnids form neck-teeth in response to invertebrate predators and longer spines and 'helmets' in response to fish predators (see Havel 1987 for a review), tadpoles develop a broader tail and red coloration with distinct black spots when exposed to odonate predators (McCollum and Leimberger 1997), and crucian carp become deeper bodied in the presence of piscivorous fish (Brönmark and Miner 1992).

2. The prey organisms have reliable *cues* of detecting the presence of the predator in the habitat so that they can produce the morphological defence before it is too late.
3. The prey *benefits* by having the induced morphology; the morphological defence should increase the probability of surviving an encounter with a predator.
4. The inducible defence structure incurs a *cost* to the prey organisms. In the absence of predators, individuals with the induced defence should suffer a fitness cost in comparison with undefended conspecifics. If there were no cost associated with the defence structure, the prey organisms might as well have the defence all the time.

Studies on freshwater organisms that have inducible defences have shown that these defences are expressed in response to water-borne chemical cues from the predators. Some cladoceran zooplankton, such as *Daphnia*, develop a longer spine when exposed to fish chemicals, whereas other species respond by growing neck-teeth in response to chemical substances released by the invertebrate predator *Chaoborus* (Fig. 4.29). Prey protozoans develop large extensions when they coexist with predatory protozoans. Most examples of inducible defences involve invertebrates, but in recent years it has been shown that vertebrates have also evolved this kind of defence adaptation. When pike were introduced to a pond without piscivores, crucian carp developed a deeper-bodied morphology over just a few months (Brönmark and Miner 1992). Later laboratory experiments confirmed that the change in morphology was in response to waterborne chemical cues associated with the diet of the predator. Chemical cues from predatory fish feeding on benthic macroinvertebrates did not elicit a change in body morphology, whereas when the predator was feeding on fish, crucian carp became deeper-bodied (Brönmark and Pettersson 1994). Tadpoles are another example of a vertebrate with an inducible defence. Odonate larvae may be important predators on tadpoles and McCollum and Leimberger (1997) showed that hylid tadpoles exposed to odonate cues developed a broader tail, which also changed in colour to bright red with distinct black spots. There are also indications that the green alga *Scenedesmus* responds to the presence of herbivores by growing larger in an attempt to become less vulnerable (Hessen and van Donk 1993; Fig. 4.29).

Costs of having an inducible defence morphology have been demonstrated in several organisms. For example, some prey have a slower growth rate and/or a reduced output of offspring in defence morphology in the absence of predators. However, the results are not always clear-cut. In daphnids, for example, food availability has been shown to be important in how, or even if, the costs are expressed. The costs of having a defence morphology should be dependent on resource availability. When resources are abundant, an organism can afford the cost, whereas when resources are limited a morphological defence structure can incur a cost. A field experiment with crucian carp showed that deep-bodied carp had the same growth rate as shallow-bodied carp when carp densities were low (resource competition weak), whereas at high densities resources became limited and then the cost of being deep-bodied was expressed as a slower growth rate (Pettersson and Brönmark 1997).

Inducible morphological defences are fascinating adaptations and deserve detailed study in themselves, but they are also interesting because they provide evolutionary biologists with an opportunity to test possible costs and benefits of various structures without being constrained by phylogeny (i.e. comparisons can be made within the same species). In most cases the costs and benefits of a specific structure have to be inferred from a comparison of different species, and then other factors that differ between the species may affect the outcome of the comparison.

Parasitism

A parasite is an organism that receives its nutrition from a host without killing it, at least not initially. Parasites can be classified into **microparasites**, including viruses, bacteria, fungi, and protozoa, and **macroparasites** such as helminth worms (e.g. tapeworms, trematodes, and flukes), nematodes, and crustaceans. Microparasites are characterized by having a small size, short generation time, and by multiplying directly within their host. Macroparasites grow in their host but need an intermediate host for reproduction. Most parasites are **endoparasitic** (i.e. they live within the body of their host). **Ectoparasites**, on the other hand, live on the surface of their host. In fish, for example, ectoparasites may attach to gill filaments or fins or directly on the body surface (e.g. fish lice and lampreys). Microparasites are often transferred directly from host to host, whereas macroparasites generally are transmitted by another species, a **vector**. Macroparasites often have very complex life cycles with intermediate hosts in other habitats (e.g. terrestrial).

Pathogenic bacteria and viruses may cause dramatic increases in the mortality rates of populations. The bacterium *Aeromonas*, for example, was suggested to cause a 98% mortality in a population of perch (Craig 1987). A number of studies have suggested that parasites may influence and even regulate plankton populations, both phytoplankton and zooplankton. Final stages of algal blooms are often infected by fungal parasites. Whether the parasite is actually causing the bloom to collapse, or whether it just infects algae that are already dying is still unclear. A study on daphnids in Lake Constance revealed a high infection (up to 50%) of a protozoan endoparasite (*Caullerya mesnili*) and laboratory experiments revealed that the parasite dramatically increased mortality and reduced fecundity of

Daphnia galeata (Bittner *et al.* 2002). In microcosms the parasite drove *Daphnia* populations to extinction within 10–12 weeks. Further, parasite infections may make *Daphnia* more conspicuous and thus more vulnerable to predation by visual predators.

However, parasites generally affect their hosts more slowly. A large fraction of the energy budget of parasite-infested individuals is diverted to the parasite, resulting in less energy being available for growth and reproduction of the host. The increased energy demand of infested individuals may also result in behavioural changes. Sticklebacks infested with the cestode worm *Schistocephalus solidus* are more hungry than uninfested sticklebacks and increase their use of foraging habitats where risk of predation is high (Milinski 1985). Parasites may also cause degradation of inner organs and it has been shown that parasite deformation of the reproductive organs in snails effectively castrates it, leaving a large part of the population unable to reproduce.

Parasites that have complex life cycles with several hosts may have elaborate adaptations and strategies that increase the probability of transfer between host organisms. Many parasites change the behaviour of their intermediate hosts so that the risk of predation by the final host increases. Parasites may reduce the locomotory ability of the intermediate host, increase or decrease activity levels, or change its habitat use to habitats where predation risk is higher. For example, roach that are infected by the tapeworm *Ligula intestinalis* prefer shallower, inshore areas where they are more at risk to predation by birds. Experimental studies showed that infected roach were less active and were swimming close to the surface exposed to bird predation (Loot *et al.* 2002). Thus, parasites may change the behaviour of their intermediate host to increase their probability of transmission to their final host. Hosts, on the other hand, may have adaptations that reduce their exposure to parasites, including avoidance of consuming infected prey or avoidance of habitats with high prevalence of parasites. Sticklebacks avoid eating parasitized copepods (Wedekind and Milinski 1996) and female frogs have been shown to avoid laying their eggs in ponds that contain snails infected by a parasitic trematode (Kiesecker and Skelly 2000). The parasite has tadpoles as their final host and infected tadpoles have reduced growth rate and survival.

Epibionts

Epibionts, which are organisms that live attached to the body surface of another organism, are an intriguing type of parasite. They are bacteria, algae, protozoans, or even small metazoans attached to a host organism, often a planktonic microcrustacean. Life for an epibiont attached to a motile host organism has some interesting consequences. If the epibiont is an alga, it will benefit from an increased access to nutrients since the nutrient-depleted zone around the cell will decrease. Similarly, heterotrophic epibionts will have their particulate food renewed by the swimming of the host organism.

If the number of epibionts is not too high, the host organism generally continues its life without problems. However, thick layers of epibionts may affect swimming speed of the host organism, which in turn affects its feeding, respiration, and escape movements from predators.

Impact of freshwater parasites on humans

The prevalence of parasites in freshwater systems may have important ramifications for humans. Fish diseases and parasites may cause catastrophic declines in natural fish populations causing serious problems for fisheries but also affecting the dynamics and structure of the whole system through cascading effects down food chains. Further, the aquaculture industry is very sensitive to outbreaks of diseases and infestations of parasites. Several parasites that have freshwater organisms as intermediate hosts may also cause great economic and medical problems. Trematodes are an important group of parasitic flatworms that include liver flukes which cause 'liver rot' in sheep and the parasite causing Bilharzia (snail-fever) in man. These parasites have complex life cycles where one or several of the intermediate hosts are freshwater organisms, such as snails or fish. As an example, we will summarize the life cycle of *Schistozoma mansoni*, a trematode parasite that causes the tropical disease Bilharzia (named after the German pathologist Theodor Bilharz).

The adult schistosome worms live in the blood system of their host, usually in the veins associated with the intestine. Eggs laid by the parasite penetrate the intestine walls and leave the host's body in faeces or urine. If the eggs fall into water they hatch and develop into ciliated larvae (miracidia) that must find a suitable snail to penetrate within a few hours. Within the snails the miracidia develop into sporocysts and after about a month so-called cercariae are produced which then leave the snail. The cercariae are free-swimming in freshwaters, but must locate and enter the definitive host within a short time period. In the final host (man) the cercariae mature to adult worms and migrate via the heart, lungs, and liver to intestinal veins. Bilharzia causes damage to the gut walls and the vessels of the liver and lungs as the eggs become trapped there. This disease is a considerable problem in many tropical countries, especially in the vicinity of man-made freshwater habitats such as irrigation canals, fish ponds, and reservoirs. Concerted efforts have been made to reduce the prevalence of Bilharzia, by a combination of different methods affecting different life stages of the parasite. Medication, education, and provision of freshwater sources for the affected human population as well as large-scale programmes to eradicate the intermediate snail host have been performed. Snail populations have been reduced by applying specific molluscicidal chemicals, by changing the freshwater environment towards less dense macrophyte beds, and by introduction of predators (fish, crustaceans) or snail competitors that are not used as intermediate hosts by the parasite.

Symbiosis

Symbiosis is broadly defined as an interaction between organisms that live together without harming one another. If the association between the two organisms is beneficial to both parties, that is, the organisms have a higher growth rate, lower mortality, or higher reproductive output when living together as compared to when living alone, the interaction is termed 'mutualism'. **Mutualism** can be either obligate, where the organisms cannot survive without being associated, or facultative where one or both organisms do better in the mutualistic relationship, but can survive and reproduce alone as well. **Commensalism** is an interaction where one of the organisms benefits from the association, whereas the other is indifferent. In general, knowledge of the prevalence and importance of mutualistic relationships in freshwater is rather scarce and relatively few studies have involved mutualistic interactions, especially when compared to the wealth of studies considering competition and predation. There are, however, some examples of mutualistic interactions in freshwaters, typically between an alga and a plant or an invertebrate.

The freshwater fern *Azolla*, which grows on the water surface, has special cavities in the leaves that house a symbiotic, nitrogen-fixing cyanobacteria (*Anabaena*). The fern host may obtain all or most of the nitrogen it needs from the cyanobacteria, whereas the fern provides its symbiont with organic substances. The symbiotic association with the nitrogen-fixing cyanobacteria results in *Azolla* having a high nitrogen content and this is utilized in, for example, Vietnam and China where *Azolla* is used as a fertilizer in rice fields.

Several freshwater invertebrates form close symbiotic relationships with algae of the genus *Chlorella*, where the algae are maintained endosymbiotically within the cells of their hosts, which may be protozoans, sponges, or hydroids. The invertebrates and algae both gain by having their nutritional processes tightly coupled. The invertebrate hosts supply the algae with nutrients (phosphorus and/or nitrogen) and carbon dioxide, whereas the invertebrates benefit from the association by being provided with organic substances produced in photosynthesis. This symbiosis makes the host partly independent of external food for survival and growth, which is a considerable competitive advantage in places where, or during periods when, food is scarce (Stabell *et al.* 2002).

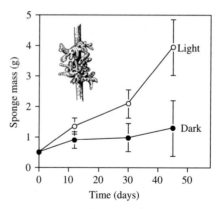

Fig. 4.30   The growth rate of the sponge (*Spongilla* sp.) when grown under dark and light conditions, demonstrating the positive effect of an algal symbiont. Modified from Frost and Williamson (1980).

Another example of symbiosis is that the growth rate of freshwater sponges depends to a large extent on the association with algae, which has been shown experimentally. If grown in complete darkness the algal symbionts die. The growth rate of sponges without algae could then be compared with normal sponges with algal symbionts and it was found that the algae increased the growth rate of the sponge from 50% to 80% (Frost and Williamson 1980; Fig. 4.30).

Practical experiments and observations

Functional response

Background   The functional response describes the consumption rate of an individual consumer in relation to increasing prey densities. The shape of the functional response may differ between consumer species and, further, defence adaptations in the prey organisms, interactions between predators, habitat complexity, and availability of alternative prey may affect the functional response.

Performance   Three-spined sticklebacks, bluegill sunfish, or even aquarium fish like guppy can be used as a predator. In the basic set-up, the consumption rate of the predator is estimated in containers with different prey densities. Introduce the fish to the container and let it acclimatize. Then introduce the prey organisms and monitor the consumption rate of the fish. The experimental time should be kept short so that feeding rate is unaffected by prey depletion and/or predator satiation. The basic set-up can be varied as follows:

1. *Different prey species.* Use different species as prey (e.g. *Daphnia*, copepods, *Chaoborous*, oligochaetes, mayflies).
2. *Differences in predator density.* Investigate how the functional response is affected by interactions between predators by doing the functional response experiment at increasing predator densities. However, be sure to follow the consumption rate of an individual fish.
3. *Differences in habitat complexity.* Increase the structural complexity of the habitat, for example, by adding plants. Artificial plants can be easily made from pieces of polypropylene rope that are tied to stone at one end and untwined at the other end.
4. *Availability of alternative prey.* Add two different preys to the experimental containers (e.g. *Daphnia* and oligochaetes), and register consumption rates. If the prey are added in different proportions you can study how relative availability affects switching between prey.

To discuss   When you have terminated the experiment, calculate the number of prey eaten per predator and unit time and plot the functional response. Are they types I, II, or III curves (Fig. 4.20)? How does alternative prey, habitat complexity, etc., affect the functional response? Why is there a change? You may repeat the experiment with a filter-feeder (e.g. a mussel or a zooplankton) or a predatory invertebrate to see if you obtain the same patterns.

Predator detection in snails

Background   The snail *Physa fontinalis* has finger-like extensions of the mantle that cover part of the shell. These extensions are sensitive to tactile stimuli. A predator that touches these extensions may trigger a sequence of behaviours in the snail, starting with a slow shaking of the shell, which then increases in frequency and amplitude if the disturbance is continued. Finally, the snail detaches itself from the substrate and floats to the surface, shaking wildly.

Performance   Investigate how different predators affect the behaviour of physid snails. Collect, in the field, snails and predators such as leeches (e.g. *Glossiphonia complanata*, *Erpobdella* sp., *Helobdella* sp.), flatworms, and larvae of the great diving beetle. Put the snails in a shallow container with pond water and let them acclimatize for a while. When they are moving around take a predator with a pair of tweezers and touch the mantle extensions of the snail with the predator. Record if the snail reacts and how much. Compare the effect of the predator with just touching the snail with a pair of tweezers. First use a clean pair of tweezers, and then tweezers that have held the predator.

To discuss Does the snail react to all predators? Why is there a difference? Is the snail reacting to mechanical or chemical stimuli?

Vertical migration

Background In lakes and ponds with fish many zooplankton, such as *Daphnia*, migrate vertically in the water column, spending the daylight hours close to the bottom and then migrating up into the water column during the night. Chemical cues from fish are suggested to trigger this behaviour. The effect of fish cues can be examined in a simple laboratory experiment. Take glass or plexiglas cylinders and cover them with aluminium foil. Add aerated tap water and *Daphnia* to the cylinders. Divide the experimental cylinders into four groups. To two groups add water from a tank with a fish that has been eating *Daphnia*, and add a similar amount of tap water to the other two. Illuminate one fish-scented and one control cylinder from above and cover the other two with a lid so that no light enters. Let them stand for a few hours and then gently remove the foil and register the vertical position of the zooplankton. Instead of *Daphnia* you could use *Chaoborous*.

To discuss What is the proximal cue that affects the migration of zooplankton? What is the value of this behaviour? What are the costs?

Food preference in crayfish

Background Crayfish are generally considered omnivorous foragers, feeding on detritus, macrophytes, periphytic algae, and benthic macroinvertebrates. Nevertheless, at high population densities they often have a strong impact on the abundance and distribution of freshwater macrophytes. However, not all macrophyte species are equally affected, possibly due to crayfish preference for certain species.

Performance Investigate how characteristics (morphology, texture, submerged versus emerged, etc.) of different macrophyte species affect crayfish feeding rates. Collect different macrophyte species, such as *Ceratophyllum*, *Elodea*, *Potamogeton*, *Sparganium*, *Typha*, and *Phragmites*, from a nearby lake or pond. Weigh the individual plants of each species and then introduce an individual of all the species into each experimental tank. Add a crayfish that has been starved for 24 h and let the experiment run overnight. Remove the crayfish, re-weigh the plants, and calculate losses due to grazing.

To discuss Which macrophyte species were preferred? What characteristics of the macrophytes made some species less preferred? Could macrophytes be affected by crayfish other than by direct grazing?

Area restricted search

Background Resources are often patchily distributed and to utilize them effectively foragers need to be able to change their foraging behaviour in response to resource levels. The effect of differences in resource levels on foraging behaviour could be investigated in a herbivore, the pond snail *Lymnaea stagnalis*.

Performance Place some clay tiles on the bottom of a container. Fill the container with lake water so that the tiles are covered. Make a periphyton inoculate by brushing off periphyton from stones and submerged macrophytes and add this to the container together with a few drops of plant nutrient. If you start a tile container every second day for a week you will obtain tiles with different periphyton biomass. Gently move a tile to an experimental aquarium and add a snail that has been starved for 24 h. Make a map of the movement patterns of the snail and note the position at regular time intervals so that you can calculate moving speed, number of turns made, distance moved, etc.

To discuss Does the snail respond to changes in resource levels by modifying foraging behaviour? What happens when the snail finds a patch (tile) of periphyton?

Algal grazer resistance

Background Some algal species are less susceptible to grazers than other. The aim of this experiment is to illustrate how zooplankton grazing may shape the algal community.

Performance Obtain water from a eutrophic lake. Remove any large zooplankton by pouring the water through a net with a mesh size of about 100 μm. Then pour the water into a jar or aquarium and add nutrients (a few drops of commercial fertilizer for potted plants works well). Put the algal culture under light and at a temperature between 18° and 25°C until the water becomes green, which generally takes one to two weeks. In the meantime, you could return to the lake and haul for zooplankton. Divide the algal culture into two aquaria and put the zooplankton in one and keep the other as a control. If you want to test differences statistically, you should replicate the two treatments at least three times. Leave the aquaria undisturbed for a few days. Take out a sample from each aquaria, preserve it with Lugol's iodine and look at the samples. Which sample has the highest abundance of algae? Which algal groups and morphologies are dominant in the control? Which are dominant in the treatment with grazers?

You can augment this experiment by initially separating the grazers into different size fractions. This can be done by gently pouring the zooplankton culture through a 200 μm mesh (mainly large *Daphnia*), then through a 100 μm mesh (mainly smaller cladocerans and copepods), and finally through a 50 μm mesh retaining copepod nauplius and rotifers.

To discuss What happens in the algal community when experiencing grazing by the different size fractions of zooplankton? Which morphological feature seems to be the best protection against grazing? Why are different algal groups dominant when grazed by different grazer size classes?

5 Food web interactions in freshwater ecosystems

Introduction

In previous chapters we discussed important aspects of the abiotic frame and presented the organisms and their interactions, for example, competition, predation, and herbivory. As indicated by the symbol for this Chapter (see above), we will now put all these features of freshwater ecosystems together into a complex synthesis and focus on the large-scale connections among and between organisms and their environment. A suitable introduction to this subject is to consider the earlier controversial opinion that organisms in a lake affect each other, and that higher trophic levels have a profound influence on the amount of algae (Andersson 1984), a view that met severe resistance when first introduced. As we have seen in Chapters 2 and 4, there is a strong relation between algal biomass (chlorophyll *a*) and phosphorus loading (e.g. Vollenweider 1968; Fig. 5.1). Furthermore, experimental addition of nutrients in small-scale enclosures as well as in whole lake manipulations (Schindler 1974), commonly results in increased algal biomass. However, a closer look at the regression between phosphorus and algal biomass reveals that a lot of the variation remains unexplained, and that for a given phosphorous concentration, the variation in chlorophyll *a* concentration may be considerable (Fig. 5.1). Moreover, the regression is based on log-transformed data, which means that low values become relatively more influential than higher values, thus reducing the variation. However, if the regression is based on untransformed data, it becomes even more obvious that the relation between phosphorus and chlorophyll *a* contains a significant amount of unexplained variation (Fig. 5.1). What then is the reason for this unexplained variation? The importance of biotic interactions for structuring freshwater communities has been recognized and many theoretical and empirical studies have suggested that differences in

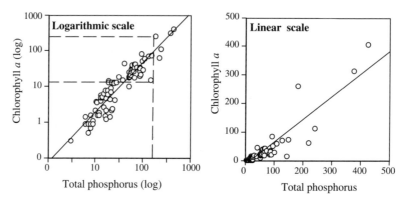

Fig. 5.1   The relation between total phosphorus and chlorophyll a in lakes. The same dataset is shown on logarithmic and linear scales, showing that the variation in chlorophyll a concentration at similar phosphorus concentration is considerable. The broken lines indicate that at a phosphorus concentration of 200 μg l⁻¹ the chlorophyll a concentration can vary between 15 and 280 μg l⁻¹!

concentrations of algal biomass at a given nutrient load is due to differences in higher trophic levels, such as zooplankton and fish (e.g. see Carpenter *et al.* 1985; Kerfoot and Sih 1987; Carpenter and Kitchell 1993).

Much of our original knowledge about complex or indirect interactions in freshwaters arose from studies carried out in pelagic habitats. However, as we shall see, the pelagic zone of a waterbody is closely linked to the more structurally complex littoral and benthic habitats, as well as to the microbial components of the food web. Before we dig deeper into these links, we will examine the intriguing roots of modern food web theory.

The evolution of an ecosystem approach

Naturalists and people spending time in natural ecosystems have always had an instinctive feeling that organisms interact with each other and with their environment. Indeed, as early as 1887, Forbes expressed the necessity for studying species interactions as well as effects of environmental conditions in order to understand the life cycle of freshwater organisms (Forbes 1925/1887). In the early history of the science of ecology there was a long-standing controversy about whether the distribution and abundance of species was limited by abiotic, density-independent, or by biotic, density-dependent processes. In 1960, Hairston, Smith, and Slobodkin formulated a conceptual model that reconciled these opposing viewpoints and highlighted the importance of trophic interactions for the structure of natural communities. The paper by Hairston *et al.* (1960) has been enormously influential, in spite of the lack of any new data, which makes it more of a philosophical reflection than a pure scientific publication. The philosophical touch was even more pronounced in the original manuscript, simply entitled: 'Étude' (French expression for 'study'). In the first section, which was omitted in the published version, the authors stated: '*The observations that will be used are, we feel, so universal as to be self-evident. Any merit that the communication may have, therefore, lies in the use that we have made of these observations.*' Indeed, the most important conclusions may be drawn from the most self-evident observations, and we will look a bit closer into how theory, observations, and experimentalism interacted in creating one of the more influential branches of modern ecology.

The theoretical take-off

It all started with the question: *Why is the world green?*, that is, why are the herbivores not able to take advantage of all, or at least a larger part of, the plant biomass produced? Hairston *et al.* (1960) argued that herbivore populations are limited by predators and, thus cannot expand to densities where they can limit plants. In food chains with three trophic levels, then, predators will be limited by the availability of resources (herbivores), herbivores limited by predators, and plants by resources, such as light and space (Fig. 5.2). Therefore, the principal factor limiting growth of a population will depend in which trophic level the population is positioned, and the dominant structuring force will flip-flop down the food chain so that predators are regulated by competition, herbivores by predation, and primary producers by competition. In this way, the well-being of green plants, be they grass, trees, or algae, is closely connected to the activities and abundances of predators, whether they are birds, wolves, or pike. Although it is a very simple model, it has become one of the more influential in ecology and has even got its own nickname, the *HSS model*, after its founders.

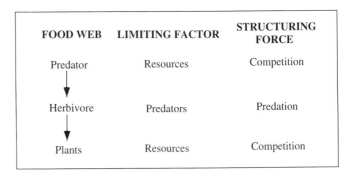

Fig. 5.2   The predictions from the HSS model with regard to what factors are limiting and the most important structuring forces at each trophic level in a food chain with three trophic levels.

Of course, their arguments were not unchallenged, and the paper provoked a number of reactions and formulations of alternative explanations for the lack of herbivore control of plants. For example, chemical defence systems (secondary plant substances) were suggested to limit the exploitation of plants by herbivores and it was questioned whether 'trophic level' really is an ecologically meaningful concept. After an initial period of intense debate, the HSS model was, if not forgotten, abandoned for other hypotheses. For a long time community ecologists mainly focused on the importance of competition for creating patterns and structure in natural communities.

The empirical beginning

Meanwhile, back in the lake, or, in fact, in fish ponds in the former Czechoslovakia, a group led by J. Hrbácek observed that when fish populations were reduced, large zooplankton (*Daphnia*) became abundant with a simultaneous decrease in algal biomass and increased water clarity. In ponds with high fish abundance, however, there was a dominance of small zooplankton species, such as *Bosmina*, and a high biomass of phytoplankton (Hrbácek *et al.* 1961). The mechanisms behind these patterns were not clear at the time, but in 1965 Brooks and Dodson presented a theory, the *size-efficiency hypothesis*, that explained the mechanisms responsible for these patterns. They argued that large cladoceran zooplankton, such as *Daphnia*, are more vulnerable to planktivorous fish and this results in a dominance of small-bodied, less vulnerable zooplankton at high fish densities. At low fish densities large zooplankton dominate because they are superior competitors; they are more efficient grazers and can feed on a wider size range of particles and, further, they have a lower metabolic demand per unit mass. The size-efficiency hypothesis predicts that when planktivory is intense, and small, inefficient zooplankton species predominate, the algal biomass will be high, whereas at low planktivore densities, large zooplankton will efficiently reduce algal biomass.

Cascading trophic interactions

Surprisingly, there was little progress in the application of food chain theory to freshwater systems during the following years. However, at the beginning of the 1980s, knowledge about food web interactions developed, following the launch of the *cascading trophic interaction* concept, which is based on the HSS model, Hrbácek's pond studies, and the size-efficiency hypothesis (Carpenter *et al.* 1985). A *trophic cascade* is an indirect interaction characteristic of linear food chains where a predator species A has an indirect positive effect on a plant species C by reducing the abundance of the herbivore species B. If the indirect positive effect on species C is mediated through a change in the behaviour of species B, the effect may be referred to as a *behavioural cascade* (Romare and Hansson 2003). The trophic cascade hypothesis partly explains the variation in primary productivity that remains unexplained by the effects of nutrient input as a result of trophic interactions in the food web. More specifically, changes in fish assemblage structure, the top trophic level, are expected to cascade down to lower trophic levels and finally affect the primary producers. For example, in a simplified food web (Fig. 5.3)

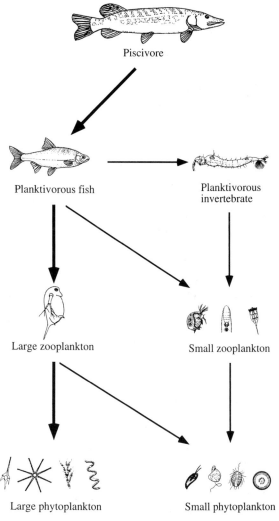

Fig. 5.3   A simplified, pelagic food web. The thickness of the arrows indicate the strength of interaction. Modified after Carpenter and Kitchell (1993).

an increase in abundance of size-selective piscivores is predicted to decrease the density of planktivorous fish and to result in an increasing importance of invertebrate predators, such as *Chaoborus*. Invertebrate planktivory and competition among zooplanktivores leads to a shift towards a dominance of larger zooplankton species (*Daphnia*). Large zooplankton can ingest a broader range of algal cells and, in addition, they have a lower rate of nutrient excretion per unit mass. This means that they will regenerate nutrients at a slower rate, resulting in a reduced biomass of phytoplankton. A reduction of piscivore density is expected to have opposite effects: increasing planktivore populations, dominance of small zooplankton species, and an increase in phytoplankton biomass. The predictions of the cascading trophic interactions concept have been tested in a large number of experimental manipulations at different scales (summarized in Carpenter and Kitchell 1993) in different parts of the world. Below, we will give some examples of empirical tests of the importance of food chain interactions in freshwater systems and also discuss critiques and developments.

Experimental illustrations of trophic cascades

Small-scale experiments

An early experiment on how higher trophic levels could affect the amount of algae and even phosphorus concentration was carried out in the eutrophic Lake Trummen, southern Sweden (Andersson 1984). Plastic enclosures put in the open water of the lake were stocked with different combinations of fish, including roach (*Rutilus rutilus*; Fig. 3.21) and bream (*Abramis brama*; Fig. 3.21). At the end of the experiment, there were considerable differences between enclosures. The density of the prey organisms *Daphnia* and chironomid larvae was lower in enclosures with fish, compared to controls without fish (Fig. 5.4). In addition to this direct effect from fish predation, the presence of fish also had a profound impact on the development of algal chlorophyll and the concentration of total phosphorus, both showing much higher concentrations in the presence of fish. Moreover, water transparency was reduced by more than 1 m when fish were present. This early experimental study elegantly illustrated that activities at higher trophic levels may well cascade down to algae and even affect the amount of nutrients present in the system. However, the experiment was carried out in plastic enclosures and although small-scale experiments have their merits, the impact of the manipulation may be exaggerated because compensatory mechanisms that may operate in lake systems are not at work at this smaller scale.

Large-scale experiments

In order to test predictions from the trophic cascade theory under more natural conditions, Carpenter and Kitchell and their co-workers have

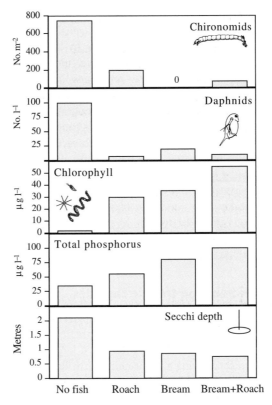

Fig. 5.4   Effects of the fish species roach (*R. rutilus*) and bream (*A. brama*) on abundance of chironomid larvae and daphnids, and concentrations of chlorophyll and total phosphorus, as well as water transparency. Data are from an enclosure experiment carried out in 1978 (Andersson 1984).

manipulated, in a series of experiments, the fish assemblage of two lakes in northern Michigan, USA, and compared the treatment effects to an untouched control lake (Carpenter and Kitchell 1993). In one of the earlier experiments they performed reciprocal fish replacement between the two experimental lakes. One of the lakes, Tuesday Lake had no piscivorous fish but high population densities of planktivorous minnows prior to the manipulation. Ninety per cent of the minnows were removed from the lake and piscivorous largemouth bass were introduced. This resulted in a dramatic shift in zooplankton community structure from a dominance of copepods, rotifers, and small cladocerans, such as *Bosmina*, to a dominance of large-bodied cladocerans. The biomass and productivity of the phytoplankton were reduced. Largemouth bass was originally the dominant fish species in the nearby Peter Lake and during the manipulation, 90% of the bass were removed and the 49 601(!) minnows from Tuesday Lake were introduced. The effects of the manipulation of the Peter Lake system were contrary to expectations. Initially, the increased planktivore population had no effect on zooplankton and phytoplankton; zooplankton was still dominated by large-bodied cladocerans (*Daphnia, Holopedium*) and the phytoplankton biomass remained at low levels. However, by the end of the summer, densities of *Daphnia* declined followed by an increasing phytoplankton biomass. It turned out that the few remaining largemouth bass had a considerable impact on the introduced minnows, which almost disappeared from the pelagic zone due to predation, emigration from the lake, and by changing habitat to the shallow water littoral zone in response to predation threat. The remaining bass reproduced and formed a strong year-class and when these young-of-the-year bass moved into the pelagic zone in midsummer and became planktivorous, they had a significant effect on the zooplankton and phytoplankton. Thus, the manipulation of fish community structure in Tuesday Lake (addition of piscivore, removal of planktivore) resulted in effects on lower trophic levels that followed predictions from the trophic cascade model, whereas behavioural effects of a few piscivores in Peter Lake resulted in effects that were contrary to the trophic cascade model.

Natural experiments on trophic cascades

Significant changes in fish assemblages due to natural causes may provide us with 'experiments' where the predictions from food chain theory can be evaluated in natural systems on a large scale. A massive fish kill in Lake Mendota during an unusually warm summer provided an example of such a natural experiment (Vanni *et al.* 1990). The dominant planktivore (cisco, *Coregonus artedii*) almost disappeared in late summer 1987, probably due to unusually high water temperatures and hypolimnetic oxygen depletion. This resulted in striking effects on zooplankton and phytoplankton. The small-bodied *Daphnia galeata* was replaced by a large-bodied species, *Daphnia pulicaria*, resulting in a dramatic increase in zooplankton grazing rates and an equally dramatic decrease in algal biomass, especially cyanobacteria. Nutrient availability and other physical factors did not change after the fish kill, indicating that the phytoplankton response was indeed due to trophic interactions.

'Bottom-up:top-down' theory

Although the cascading trophic interaction model proved to be quite successful in a number of studies, the concept was not embraced by all researchers. Parallel to the trophic cascade theory, the *bottom-up:top-down* concept evolved (McQueen *et al.* 1986). This concept combines the predicted influences of both predators (fish; from the top of the food chain and downwards: top-down) and resource availability (from the bottom of the food chain and upwards: bottom-up) by predicting that nutrient availability determines the maximum attainable biomass, but that realized biomass is determined by the combined effects of bottom-up and top-down forces.

In contrast to the cascading trophic interaction theory, the 'bottom-up: top-down' concept argues that trophic levels near the base of the food chain will be primarily affected by bottom-up processes, whereas these effects will weaken higher up the chain. Conversely, top-down effects should be significant at the top of the chain and be of minor importance for lower trophic levels. Further, the 'bottom-up:top-down' concept predicts that at high nutrient concentrations there will be no influence on algae from fish, since the system will mainly be governed by bottom-up forces. Hence, only in low productivity (oligotrophic) systems will the effect from fish be important for the development of the algal community.

For some time there was an animated debate concerning which factors determine the structure and dynamics of lake food chains and the most extreme researchers argued that *only* top-down, or, at the other extreme, *only* bottom-up forces, were important. Nowadays, most freshwater scientists agree that both nutrients and the food web composition are important for the observed structure of the food chain in a specific lake ecosystem. Hence, instead of discussing which concept is the correct one, it is now more a question of *when*, *where*, and *how much* biotic interactions, such as predation and grazing, and abiotic constraints, such as nutrient availability, are important.

Further theoretical developments

The HSS model was originally developed for terrestrial systems and for food chains with only three trophic levels. However, Fretwell (1977) and Oksanen *et al.* (1981) developed a theoretical model that analyses the effects of potential productivity of an ecosystem on the number of trophic levels and equilibrium biomass at each trophic level, the *ecosystem exploitation hypothesis*. They argued that the number of trophic levels increases with increasing potential productivity of a system, which in lakes may be equivalent to nutrient load (Fig. 5.5). Thus, the Fretwell–Oksanen model predicts that systems with very low nutrient availability can only sustain primary producers, such as phytoplankton. Increasing the nutrient load should result in increasing phytoplankton biomass up to a level where zooplankton populations can be sustained. At further increases in productivity the biomass of zooplankton should increase whereas the increasing grazing pressure should keep the biomass of phytoplankton at a constant level. Next, the predators (e.g. planktivorous fish) start to invade the system and at some point in the productivity gradient the planktivore populations will control the zooplankton biomass, resulting in no further increase in zooplankton with increasing productivity. As a consequence of the relaxed grazing pressure, phytoplankton biomass will start to increase again with increasing potential productivity. The last step in the model involves the addition of a fourth trophic level, the piscivores. Piscivores will keep planktivores at a constant level in this

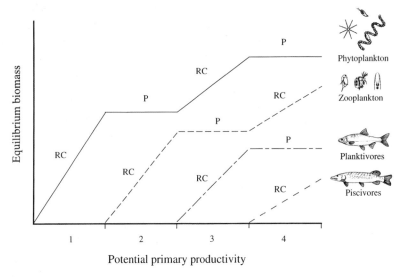

Fig. 5.5   The relation among equilibrium biomass of phytoplankton, zooplankton, planktivores, and piscivores and increasing potential primary productivity in food chains of increasing length (1–4 trophic levels). RC, resource competition; P, predation/grazing. The figure is based on the *ecosystem exploitation hypothesis* (Oksanen *et al.* 1981). Modified from Mittelbach *et al.* (1988).

region of the productivity gradient and this will result in increasing zooplankton and constant phytoplankton biomass.

Another prediction that follows is that the importance of the different structuring forces, such as competition and predation, will depend on the number of trophic levels (Persson *et al.* 1988). Phytoplankton in lakes of low productivity (low nutrient concentration) where only one trophic level can exist should be regulated by competition for resources (RC) (Fig. 5.5). When zooplankton invade the system they will be regulated by competition for food (algae) and the algae will suddenly find themselves regulated, not by resources, but by predation/grazing (P) by zooplankton. This pattern will continue when further trophic levels are added with increasing productivity; the highest trophic level will always be limited by resource competition (RC) and the next level by predation/grazing from the level above (Fig. 5.5).

Studies in European shallow eutrophic lakes have shown that at very high nutrient load, a system with four trophic levels may return to a system with only three functional trophic levels, dominated by planktivorous cyprinids (Persson *et al.* 1988). This may be due to several processes, including the ability of roach to exploit an alternative food source present in these lakes, cyanobacteria. Moreover, competitive interactions between juvenile roach and perch result in poor recruitment of perch to the piscivorous stage, resulting in reduced predation pressure allowing roach to become even more abundant. The high number of planktivores (both perch and roach) leads to a reduced biomass of large, efficient grazers, resulting in an increased biomass of phytoplankton. Other visually hunting predatory fish, such as pike and muskellunge, will have problems in catching their prey due to the high algal turbidity. In these lakes, the proportion of piscivore fish is often less than 10% of total fish biomass and piscivores are no longer functionally important in regulating the abundance of planktivorous fish, and the lake will return to a three-trophic level system.

Alternation between competition and predation along productivity gradients

As we saw in the previous section, the Fretwell–Oksanen model predicts that the increase in biomass along a productivity gradient follows a stepwise pattern, where each step is characterized by the addition of another trophic level. Phytoplankton biomass will be directly related to primary productivity in food chains with odd numbers (one, three), whereas in food chains with even numbers there will be no relation between productivity and biomass. The question then is whether natural systems really behave according to these predictions. The predictions could be tested either with correlational studies (i.e. analysis of equilibrium biomass in lakes with different nutrient loads), or in experimental studies where changes in biomass as a function of food chain manipulations and/or nutrient enrichment are recorded. Predictions from food chain theory may, however, be difficult to test in natural systems since increased productivity leads to higher trophic levels invading from surrounding lakes or ponds. This means that a productivity gradient for lakes with only two trophic levels is difficult to determine.

Antarctic lakes, however, lack fish due to biogeographical reasons, and are systems with two trophic levels. Although most of them are melt water lakes with extremely low productivity, the ones closest to the seashore may be very productive due to frequent visits by seals and birds using the lakes as water closets and swimming pools! Hence, lakes may range in productivity from ultraoligotrophic to highly productive, but still have only two trophic levels (algae and zooplankton). By comparing temperate three-level lakes with Antarctic two-level lakes of different productivity (here expressed as total phosphorus), predictions from food chain theory can be tested. We would expect the phytoplankton biomass in the three-level lakes to increase with increased nutrient availability. In two-level lakes, however, the phytoplankton biomass is not expected to increase, since new algal cells will be grazed by an unregulated grazer community (Fig. 5.6, Prediction). What happened in real lakes is as follows. In the three-level lakes phytoplankton biomass increased in biomass as predicted, whereas the algae in the two-level Antarctic lakes did not respond exactly as predicted. Here, phytoplankton biomass showed a slight increase with increasing productivity, but

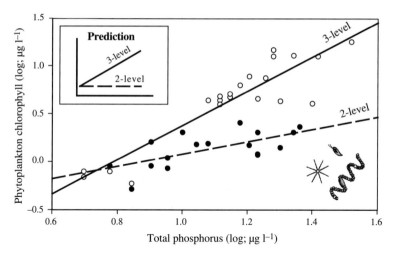

Fig. 5.6   The relation between total phosphorus and phytoplankton chlorophyll in lakes with two trophic levels (Antarctic lakes lacking fish and functionally important invertebrate preda- tors) and with three trophic levels (temperate and sub-Arctic lakes). The inset figure shows the predicted relations based on food web theory. Data from Hansson (1992*b*).

the regression slope was less steep than for three-level lakes, suggesting that at a certain phosphorus value, the algal biomass was lower in Antarctic lakes than in temperate three trophic-level lakes. That the slope of the two-level line was steeper than predicted is a result of algal responses to increased resource availability (Hansson 1992*b*). On the other hand, the regression slope was not similar to that for three-level lakes, but was forced down by grazing from an unregulated zooplankton community. Hence, the conclu- sion is that both predation/grazing and nutrients are important for the development of algal biomass in lakes and ponds.

A similar example clearly showing that the amount of algae in the water is not simply a mirror of phosphorus concentration is the study of Cockshoot Broad, England (Moss *et al.* 1996). During restoration of the Broad, the lake turned out to have low amounts of *Daphnia* in some years ('low *Daphnia* years'), but other years were 'high *Daphnia* years'. The amount of phos- phorus, as well as algal abundance varied considerably and during 'low *Daphnia* years' the chlorophyll concentration (algal abundance) was in pro- portion to phosphorus concentration (Fig. 5.7), exactly as predicted from the relation between phosphorus and chlorophyll (Fig. 5.1). However, in 'high *Daphnia* years', the chlorophyll concentrations were always around 20 μg l^{-1}, despite the fact that the phosphorus concentration varied between 50 and 170 μg l^{-1}! A plausible explanation is that at high *Daphnia* levels, any increase in algal biomass was transformed into grazer biomass, suggesting that if large zooplankton are allowed to flourish, they are able to keep algae at low levels even when nutrient availability is high.

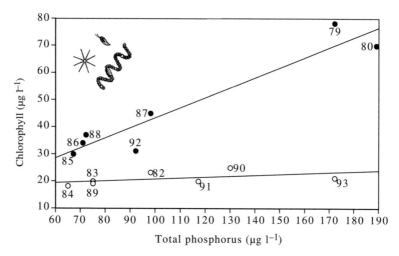

Fig. 5.7   The relation between total phosphorus and chlorophyll in Cockshoot Broad, England, showing years with high (open circles) and low (closed circles) zooplankton biomass. The figure illustrates that in the same lake, and at the same phosphorus concentration, the algal biomass can vary widely depending on the abundance of herbivorous zooplankton. (Data from Moss *et al.* 1996.)

Complex interactions in benthic food chains

As we have seen, the importance of complex trophic interactions has been demonstrated in a number of empirical studies on pelagic food chains in lakes. The predominance of studies on cascading effects in open water may in part be because this habitat is easy to sample, the communities are relatively simple, and perhaps most importantly, limnologists have a long tradition of working in this habitat (Lodge *et al.* 1988). However, the pelagic zone is not a closed system but is dependent on functional links, especially with the littoral zone within the lake, but also through interactions with the surrounding catchment area. The littoral zone is typically very patchy with a high degree of structural complexity. It has been argued that strength of complex interactions should be reduced in habitats with high complexity, but a number of studies have shown that manipulations of high trophic levels may cause trophic cascades in benthic food chains as well. Brönmark *et al.* (1992) manipulated the density of pumpkinseed sunfish (*Lepomis gibbosus*) in cages in two lakes in northern Wisconsin, USA, and estimated the direct and indirect effects down the food chain. Pumpkinseeds have strong jaw muscles and pharyngeal teeth which enable them to crush snail shells, and a large proportion of their diet (>80%) usually consists of snails. In enclosures with pumpkinseeds the biomass of snails was dramatically reduced and there was a change in species composition from large, thin-shelled snail species towards small, thick-shelled species. The decrease in snail density resulted in an increased accumulation of periphytic algae, the major food of snails. Thus, manipulation of the density of the molluscivorous pumpkinseed sunfish cascaded all the way down to the periphytic algae (Fig. 5.8). Similar results were obtained by Martin *et al.* (1994) in a study on redear sunfish (*Lepomis microlophus*) and by Brönmark (1994) using tench (*Tinca tinca*). In these studies there was also an increase in the growth of submerged macrophytes (*Najas* sp. and *Elodea canadensis*, respectively) in cages without molluscivorous fish. This was probably due to reduced shading by periphytic algae, and possibly reduced competition for nutrients, as snails decreased the thickness of the periphyton layer on submerged macrophytes. Laboratory studies have also shown that periphytic algae decrease the availability of light for submerged macrophytes (Sand-Jensen and Borum 1984), and that grazing snails indirectly may

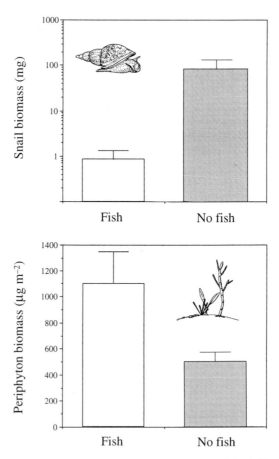

Fig. 5.8 The biomass of snails and periphyton on artificial substrates (plastic flagging tape) in cages with and without pumpkinseed sunfish. Modified from Brönmark *et al.* (1992).

increase the growth rate of submerged macrophytes by removing the algal cover (Brönmark 1985).

Direct and indirect effects of benthic omnivores

Crayfish are large and, in many systems, common benthic feeders that could be expected to affect interactions in benthic food chains. However, crayfish are **omnivorous** foragers (i.e. consume resources from different trophic levels). Macrophytes, periphyton, benthic macroinvertebrates, and detritus are all eaten by crayfish, and, in addition, they are often found scavenging on dead fish. Where omnivores are prominent, the significant direct effects down food chains may be counteracted by indirect effects, thus reducing interaction strength. Thus, it is not intuitively obvious how experimental manipulations of crayfish densities would affect lower trophic levels. For example, in analogy with the studies on snail-eating sunfish and tench described above, crayfish could be expected to reduce snail density, increase the biomass of periphyton, and then reduce the growth of submerged macrophytes. On the other hand, crayfish could have negative effects on both macrophytes and periphyton through direct consumption (Fig. 5.9). Lodge *et al.* (1994) tested the significance of direct and indirect interactions by manipulating the population density of a crayfish, *Orconectes rusticus*, in experimental cages placed in a lake. They found that it had an appreciable negative effect on snails and submerged macrophytes, whereas the biomass of periphyton increased. The reduction in submerged macrophytes was attributed to direct consumption by crayfish rather than indirect effects of an increasing periphyton cover. Similar results were found in a Swedish study where the indigenous noble crayfish and the exotic signal crayfish had significant negative effects on snails and the submerged macrophyte *Chara hispida*, whereas there was a positive, indirect effect on periphyton biomass (Nyström *et al.* 1999).

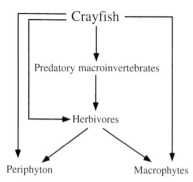

Fig. 5.9   Interactions in a food web with crayfish as the top predator.

Behaviourally mediated interactions

In most studies showing strong cascading effects in freshwater food chains, these have been caused by direct, lethal effects of predators. However, predators also affect the *behaviour* of their prey resulting in reduced activity or a change in habitat/refuge use. Predator-mediated changes in the habitat use of an intermediate consumer may affect its diet and thus change the strength and direction of interactions in food chains. For example, presence of the piscivorous largemouth bass (*Micropterus salmoides*; Fig. 3.21) affects the habitat use of juvenile sunfish (*Lepomis* spp.; Fig. 3.21). In the absence of bass, juvenile sunfish prefer the pelagic zone, but when bass are introduced they change habitat to the littoral zone with vegetation, where they feed on benthic macroinvertebrates (Werner *et al.* 1983). Moreover, juvenile sunfish have a strong negative effect on the density of large invertebrates, resulting in a smaller average size of macroinvertebrates (Mittelbach 1988). Thus, presence of a piscivore may have significant indirect effects on the benthic macroinvertebrates in the littoral zone through a behavioural shift in habitat use of the intermediate consumer, in this case the juvenile sunfish. Similarly, Turner (1996, 1997) found that the presence of molluscivorous pumpkinseed sunfish affected the interaction between herbivorous snails and periphytic algae. Snails in lakes with pumpkinseeds spent more time in refuge from predators than in lakes without pumpkinseeds, and a laboratory experiment showed that snails changed their habitat preference in response to chemical cues from conspecifics. Turner then manipulated the perceived predation risk by adding different quantities of crushed snails to experimental pools. He found that refuge use increased in response to predation risk and, further, that this had positive effects

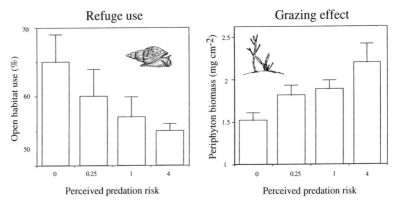

Fig. 5.10   The effect of chemical cues from pumpkinseed sunfish on the refuge use of a snail and the indirect, behavioural cascade effect on periphyton. From Turner (1997).

on periphyton biomass outside the refuge (Fig. 5.10). Thus, a change in snail behaviour mediated by chemical cues released from a predator had strong effects on the primary producer!

Alternative stable states

Ecosystems exposed to gradually changing environmental conditions, such as climate, nutrient loading, or pollution, may respond in different ways. Some ecosystems respond by a gradual change (Fig. 5.11(a)), whereas others seem to be able to occur in two alternative states over a range of environmental conditions. Theoretical models have for long predicted that natural systems may occur in *alternative stable states* and that the shift between states is abrupt, not gradual (e.g. May 1977). Recently, a number of empirical studies have accumulated confirming the presence of alternative

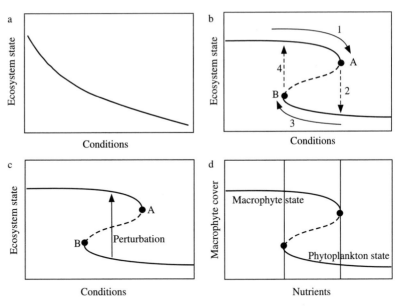

Fig. 5.11   Changes in ecosystem state in response to changes in environmental conditions. (a) An example of an ecosystem where changes in environmental conditions result in a gradual change in ecosystem state; (b) an example of an ecosystem that can exist in two alternative stable states. A change in environmental conditions (arrows 1 and 3) results in little change of ecosystem state until a bifurcation point (A and B) is reached, but then the system changes rapidly to the other state (arrows 2 and 4); (c) the ecosystem may change from one state to the other following a strong perturbation; (d) alternative stable states in shallow, eutrophic lakes (macrophyte dominance versus phytoplankton dominance). In the area between the two vertical lines the ecosystem may exist in either of the two states. Modified after Scheffer and Carpenter (2003).

stable states in very different ecosystems, from deserts to coral reefs (Scheffer and Carpenter 2003). An ecosystem may occur in alternative stable states when the response of the system is not continuous with respect to gradual changes in an environmental variable, but where the response curve instead is 'folded' backwards (Fig. 5.11(b)). An increase in the environmental variable from very low levels (arrow 1) results in little change in the ecosystem state until a threshold level, a bifurcation point, is reached (A). Then the system changes abruptly to the other state and will remain there with further increases in the environmental variable. Another important feature of such systems is that restoration of the environment back to the conditions where the transition occurred (at A) is not enough to change the system back to its original state. Instead, the system has to be reversed (3) to another bifurcation point (B) before a switch back (4) is possible. Thus, the system may occur in either of the two alternative states over a range of the environmental variable (Fig. 5.11). The stability of the alternative states is due to different buffering mechanisms inherent in the respective states. However, a sufficiently severe perturbation of the system may cause a transition to the other state (Fig. 5.11(c)).

Alternative states in shallow, eutrophic lakes

One of the best examples of abrupt shifts between alternative stable states comes from the study of shallow lakes subject to increases in nutrient loading (e.g. Jeppesen *et al.* 1998; Scheffer and Carpenter 2003). Shallow lakes (lakes that have complete mixing also during the summer period) may exist in two alternative states over a range of phosphorus levels (25–1000 μg l^{-1}; Moss *et al.* 1996; Fig. 5.11(d)). Historical data from two shallow lakes in southern Sweden: Lake Tåkern and Lake Krankesjön (Blindow *et al.* 1998) show that these lakes have regularly shifted between clear and turbid states at least since the beginning of the twentieth century (Fig. 5.12). In pristine conditions with low nutrient loadings, shallow lakes have clear water, large zooplankton, low phytoplankton biomass, extensive beds of submerged macrophytes, and a balanced fish community with a high proportion of piscivores. As nutrient load increases the lake may abruptly shift to its

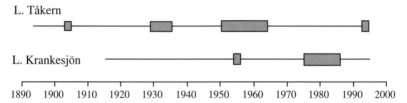

Fig. 5.12   Natural long-term shifts between clear water (thin lines) and turbid (boxes) states in two mesotrophic lakes, Lake Tåkern and Lake Krankesjön, southern Sweden. Data from Blindow *et al.* (1998).

second state characterized by turbid water where submerged macrophytes are absent, phytoplankton biomass high and often dominated by inedible cyanobacteria, zooplankton are small and sparse, and the fish community is dominated by benthivorous and planktivorous cyprinid fishes. The two states are buffered against changes by a number of different, more or less complex, feedback mechanisms. In the clear water state, macrophytes buffer against changes by reducing wind-induced sediment resuspension, 'luxury uptake' of nutrients, release of organic substances that suppress algal growth (allelopathic substances), shedding of periphyton-covered leaves, and harbouring of grazers. The turbid, phytoplankton-dominated state, on the other hand, may be buffered through shading of macrophyte shoots in early spring, absence of refuges for herbivores in the unstructured environment, resuspension of sediments by benthivorous fishes and/or a dominance of large, inedible algal species.

What causes the transitions?

The turbid state is very stable at high nutrient loads and a considerable effort is needed to 'push' the system over to the clear water state. As the clear water state is preferred from a human perspective, large resources have been invested to reverse turbid lakes to the clear water state. A mere reduction of nutrient levels is not sufficient, but the system needs a strong perturbation to switch between states, for example, a reduction in the biomass of cyprinid fishes (biomanipulation; Hansson *et al.* 1998b; see also Chapter 6). Reduced planktivory results in increased zooplankton grazing and reduced phytoplankton biomass and removal of benthivorous fishes decrease sediment resuspension and macrophyte uprooting. All this facilitates macrophyte re-establishment. The mechanisms behind the transition from the macrophyte-dominated to the phytoplankton-dominated state are less well understood. Hargeby *et al.* (2004) suggested that strong winds, causing resuspension and turbidity in the water, in combination with low temperatures during spring may delay macrophyte establishment and thereby allow phytoplankton to flourish. In a conceptual model, Brönmark and Weisner (1992) suggested that heavy mortality of piscivorous fish caused by stochastic perturbation events, such as winterkill, may change the fish community towards a dominance of planktivorous and benthivorous species. Such changes at the top-predator levels may have strong indirect effects at lower trophic levels in the food chain (see above). In the pelagic food chain this may result in increasing phytoplankton biomass and less light reaching the macrophytes. More importantly, in benthic food chains an increase in benthivorous fish should result in reduced density of grazing macroinvertebrates, especially snails, which results in an increase of epiphytic algae and a decrease of macrophyte growth due to increased shading and competition for nutrients. Experimental studies have shown that snail grazing reduces the epiphytic algal layer and promotes macrophyte growth (Brönmark 1985) and, further, that increases in the density of molluscivorous fish cause a reduction of macrophyte biomass through indirect effects in the benthic food chain (Martin *et al.* 1994; Brönmark 1994). To test the predictions from the model and the experimental studies in natural systems, Jones and Sayer (2003) did a survey of shallow lakes along gradients of fish density and nutrient concentration. Macrophyte biomass was negatively related to periphyton but showed no relationship with either phytoplankton biomass or nutrient concentration. Nutrients had no effect on periphyton biomass either, whereas increasing densities of macroinvertebrate herbivores resulted in decreasing periphyton biomass. Finally, the only factor that correlated with macroinvertebrate herbivores was fish density and, consequently, periphyton biomass was positively correlated to fish density and negatively to macrophyte biomass. Thus, these results from natural systems, as well as the experimental data, support the conceptual model in which fish determine macrophyte abundance through cascading trophic interactions down the benthic food chain, under a range of nutrient concentrations.

The marble-in-a-cup model

The alternative states and shifts between them could also be illustrated with the 'marble-in-a-cup' model (Scheffer 1990; Fig. 5.13). Here, the marble

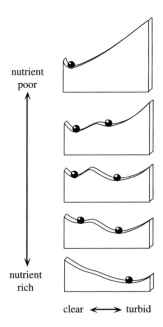

clear ⟷ turbid

Fig. 5.13   The 'marble-in-a-cup' model illustrating alternative stable states (minima) at five differ-ent nutrient levels. In very nutrient-poor conditions, the marble (i.e. a lake or pond) can only occur in a clear state. At somewhat higher nutrient levels, however, two alternative stable states, one clear and one turbid can occur. At very high nutrient levels, the lake or pond can only occur in a turbid state. From Scheffer (1990).

represents the lake, which at high nutrient concentrations lies steady in the turbid, phytoplankton-dominated state. If the marble, unexpectedly, could be moved it would most probably roll back into the turbid state (Fig. 5.13, bottom). If the nutrient concentration becomes a little lower, the two alternative stable states can be visualized as two depressions on either side of a 'hill'. The marble (lake), may now be pushed over into the clear water state, but will easily roll back again. If the nutrient concentration is reduced even more, the lake is easier to push from the turbid state and the probability that it stays in the clear water state is higher. Finally, at very low nutrient concentrations, the lake can only occur in the clear water state (Fig. 5.13, top). The model points out several interesting aspects, including: (1) not only food chain effects (fish predation on zooplankton followed by zooplankton grazing on algae), but also macrophytes are important actors in the shift between states; and (2) a low nutrient concentration increases the odds that the lake can be pushed into the clear water state.

Microbial components

When discussing ecosystem dynamics, and especially food web theory, the microbial components, including bacteria, ciliates, and heterotrophic flagellates, are seldom included. A major reason for this is that methods for quantifying processes and studying interactions at this level have not been available until recently. Despite this, the impact of microbial organisms is, as we shall see, considerable. Microbial communities differ from those formed by larger organisms mainly with respect to time and space scales, that is, everything happens during seconds instead of hours and days, and distances are not measured in metres but in micrometres. Otherwise, the organisms suffer from similar environmental constraints and are involved in interactions similar to those of larger organisms. A major breakthrough for microbial ecology in aquatic science was the identification of the *microbial loop* as being a part of the traditional food web.

The microbial loop

When trying to quantify the amount of energy (carbon) transferred in pelagic food webs, it was observed that in many lakes the carbon produced by phytoplankton was not enough to explain the growth capacity of animals higher up the food web. One of the major missing links in the attempt to quantify energy flow was the microbial community, consisting of bacteria, heterotrophic flagellates, and ciliates. These organisms occur in very high numbers in water, but they were not included in the traditional food web, which mainly focuses on algae, zooplankton, and fish (Fig. 5.3). Soon after interest had been focused on microbial components, it was discovered that bacteria retrieved part of their carbon from phytoplankton, zooplankton, and fish excreta in the form of dissolved organic carbon (DOC). Since bacteria are eaten by heterotrophic flagellates and ciliates, which in turn are eaten by macrozooplankton, rotifers, and even small fish, part of the carbon lost from the pelagic system is returned to the traditional food web by microbial components. This recycling of carbon, which was thought to be lost from the pelagic system, is termed the 'microbial loop' and clarified the majority of problems concerning the energy budget in lakes (Fig. 5.14).

Algal exudates are particularly important as a carbon source for bacteria during summer, when production is high, and at the end of algal blooms, when senescent algae release high levels of carbon. Up to 50% of the carbon fixed in algal photosynthesis is exuded (released) and used by bacteria, suggesting that bacteria receive a major part of the carbon from algae. Also, excretion products from higher trophic levels, such as fish and zooplankton, provide important inputs to the microbial loop (Fig. 5.14).

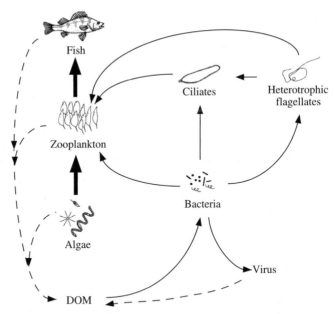

Fig. 5.14   Carbon and nutrient transport in the microbial loop and its connections with the conventional food chain, including algae, zooplankton, and fish (thick arrows). Bacteria consume dissolved organic material (DOM) excreted from organisms or supplied from the catchment area. Bacteria are then eaten by heterotrophic flagellates and zooplankton, whereas ciliates eat heterotrophic flagellates and are themselves a prey of zooplankton. Moreover, viruses infect for example, bacteria and make them release carbon and nutrients. In this way, microbial components constitute a loop for recycling of carbon and nutrients to higher trophic levels.

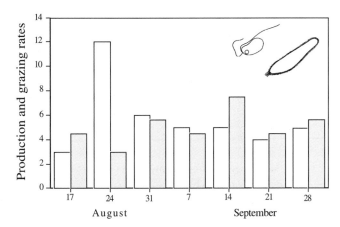

Fig. 5.15   Mean estimated bacterial production (open bars) and grazing rates by heterotrophic flagellates and ciliates on the bacteria (filled bars) in the epilimnion of the stratified Lake
Vechten, The Netherlands. The estimated grazing rates are often higher than production rates of bacteria. The units for both grazing rate and production is 10^6 bacteria
$l^{-1} h^{-1}$. Data from Bloem *et al.* (1989*b*).

Interactions among microbial components

Heterotrophic flagellates feed mainly on bacteria which are ingested
in amounts ranging between 100 and 1000 cells per individual per day,
depending on species, temperature, and bacterial abundance. The high
grazing pressure leads to bacterial abundance in the water being generally
regulated by flagellate grazing, which removes as much as 30–100% of the
bacterial production per day! The total grazing pressure by protozoans
(heterotrophic flagellates and ciliates) may thus, at times, exceed bacterial
production, as illustrated by the comparison between grazing and bacterial
production during autumn in Lake Vechten (The Netherlands, Bloem *et al.*
1989*b*; Fig. 5.15).

Ciliates feed mainly on algae and heterotrophic and autotrophic flagellates,
but many ciliates also feed directly on bacteria (Fig. 5.14). A ciliate feeds on
bacteria at about the same rate as heterotrophic flagellates, about 100–1000
cells per individual per day. However, since the abundance of ciliates is generally 20-fold lower, their impact on bacterial abundance is generally of
minor importance compared to that of heterotrophic flagellates.

Regeneration of nutrients

Bacteria

When dead phytoplankton and other decaying particles are attacked by bacteria, only low amounts of nutrients are released. This is because bacteria are

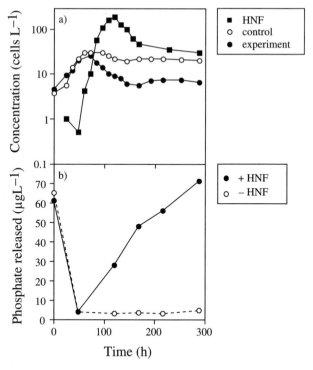

Fig. 5.16 (a) Development of bacteria in controls and in the presence of grazing heterotrophic nanoflagellates (HNF). The concentration of HNF is also given (filled squares); (b) regeneration of phosphate by heterotrophic flagellates grazing on bacteria (filled symbols) in a laboratory experiment. In the control treatment without flagellates (open symbols), the phosphate was incorporated in new bacterial biomass and not transferred further up the food chain. Data from Bloem *et al.* (1989a).

primarily investing the nutrients into new biomass. Soon, however, heterotrophic flagellates start grazing on the bacteria, thereby releasing high amounts of nutrients bound in bacterial biomass. This can be illustrated by an experiment using water with and without heterotrophic flagellates (Fig. 5.16). In experimental chambers where bacteria were not grazed, no phosphate (PO_4) was released, whereas in the presence of heterotrophic flagellates, high amounts of phosphate were released. In cases where phosphorus occurs in high enough amounts to satisfy growth, bacteria will, however, release the phosphorus to the water without being grazed.

Consumer release of nutrients—indirect 'bottom-up' effects

As illustrated by the microbial loop, higher organisms excrete DOC which is used as a substrate by bacteria. In the same way, nutrients such as phosphorus and nitrogen, are excreted in large amounts by consumers, especially larger ones such as fish (Vanni 1996). Hence, when planktivorous fish become abundant they not only reduce the grazing pressure from zoo-plankton by predation ('top-down' effect), they will also promote phyto-plankton growth by increasing the amount of phosphorus available (indirect 'bottom-up' effect).

Excretion, and thereby recycling of nutrients by grazers and predators is especially important when algae and bacteria are nutrient-limited. If a grazer prefers, for example, the diatom *Cyclotella* as food, it will excrete nutrients after digesting the meal, which are used by the less preferable alga *Aphanizomenon*. In this way consumption may affect species composition in the prey assemblage not only by reducing the number of a certain prey, but also by providing resources for a competing, less preferable, prey! Nutrients are usually excreted in easily accessible forms, such as phosphate and ammonia, which can be rapidly taken up by bacteria and algae. On a short timescale, excretion by *Daphnia* is more important than excretion by copepods, since *Daphnia* lives in a perennial state of diarrhoea. Copepods instead form faecal pellets which may sink to the sediment surface before nutrients are slowly dissolved from the pellet.

Interaction strength in freshwater food chains

The general application of the cascading trophic interaction concept has been very successful but it has nevertheless been subject to critique (Strong 1992; Polis and Strong 1997). In a review, Strong (1992) concluded that effects of trophic cascades were only prominent in species-poor communit-ies with a few strong interactors, **keystone species**, at the predator and herbi-vore level, and with algae at the bottom of the food chain. In systems with more species, the strong cascading effects would be buffered by compen-satory *shifts in species composition, temporal and spatial heterogeneity*, and *omnivory*. Other factors, such as *subsidiary inputs* from other habitats or from the detritus chain and *prey defence* adaptations, could also affect the strength of cascading interactions in food chains in general (Polis and Strong 1997). Below we bring out factors that may be important for inter-action strength in freshwater food chains. We will begin with the funda-mental question that if food webs are as complex as outlined above, can we then really talk about distinct trophic levels, or are these only artificial constructions that have nothing to do with natural, complex ecosystems?

Do trophic levels exist?

Food chain theory is based on the idea that organisms in a system can be categorized into trophic levels and that organisms at a specific trophic level feed on the trophic level below, and in turn are fed upon by organisms in the trophic level above (i.e. food chains are linear, Fig. 5.17). The patterns that we

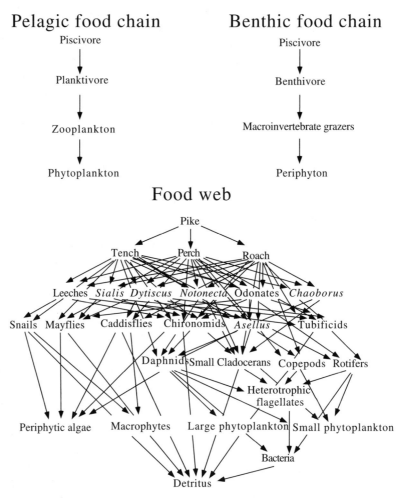

Pelagic food chain

Piscivore

↓

Planktivore

↓

Zooplankton

↓

Phytoplankton

Benthic food chain

Piscivore

↓

Benthivore

↓

Macroinvertebrate grazers

↓

Periphyton

Food web

Fig. 5.17 Two idealized, linear food chains from a pelagic and benthic habitat, and a simplified, reticulate food web.

see in the real world are naturally not as clear-cut as we may wish or as assumed by theory. Even a schematic figure of feeding links between organisms in a relatively simple pond community shows a tremendous number of interactions and a very high complexity (Fig. 5.17). Obviously, these freshwater organisms do not aggregate nicely into discrete, homogeneous trophic levels and, much less, into linear food chains. Rather, the interactions show a reticulate pattern with many consumer organisms linked to a variety of resources at different levels, from primary producers to predators or detritivores—a *food web*. After investigating the high complexity even in a simple food web of a small pond, it may seem hard to envisage that one could make any generalizations about community structure, dominant structuring forces and, even more remote, predictions of what will happen to a community when some of its components are manipulated. However, these circumstances are not as discouraging as they might seem at first glance. An interaction food web, as shown in Fig. 5.17, is based on feeding links between organisms but these say nothing about the strength of the links. It may suffice that a chironomid is found once in the gut content of an *Asellus* to establish a feeding link between these two organisms, although the diet of *Asellus* is completely dominated by algae and detritus. Thus, not all interactions are equally important in the real world, but, rather, there is a great difference in interaction strength. Further, many species can be grouped into **functional groups** or **guilds** that use the same kind of resources and react similarly to changes in the environment. By collapsing the original food web to interactions between groups of strong interactors, the systems become more manageable for analysis and for making predictions, for example, on the effects of disturbances acting on specific groups (Fig. 5.17).

When we are simplifying food webs to food chains we assign each organism to a specific trophic level; primary producers, herbivores, predators, etc. However, omnivory and other complexities in the food web results in that an organisms true position in the food web falls somewhere in between the discrete trophic levels. Stable isotope analysis provides a method to assign an organism to a trophic position, that is, a measure on a continuous scale (Fry 1991). The enrichment of $\delta^{15}N$ in food chains with 3–4 per mille at each consumer–resource step is used to determine an organism's trophic position. Thus, in simple food chains with linear consumer–resource interactions organisms in adjacent trophic levels should be separated by 3–4 per mille in $\delta^{15}N$. However, complex food web interactions obscure this picture. Omnivory (see below), for example, where consumers assimilate food from different trophic levels, may result in $\delta^{15}N$ enrichment lower than 3 per mille, and, thus, the trophic position as determined by stable isotope analysis reflects an energy-weighted average of all the trophic transfers an organism has been involved in. Vander Zanden *et al.* (1999) studied the trophic structure of food chains in a set of Canadian lakes which all had lake trout as the top predator. The lakes were classified into three categories based on the presence/absence of intermediate consumer species. The first category (A) contained lakes that were classified as having three trophic levels in a simple linear food chain, the second category (B) lakes with four trophic levels, and the third (C) lakes with five trophic levels. Trophic positions obtained from a stable isotope analysis allowed the construction of a more fine-scaled and reticulate food web model where species were grouped into trophic guilds bases on their trophic positions (Fig. 5.18). This showed that food chain length based on trophic position varied widely among lakes within each of the lake categories based on presumed food chain length. In category A, presumed to be three-trophic level lakes, food chain length varied from 3.0 to 4.8, in lakes of category B from 3.8 to 4.4, and in C lakes from 4.3 to 4.6. The stable isotope analysis also allowed a determination of the strength of interactions in the food web based on relative energy

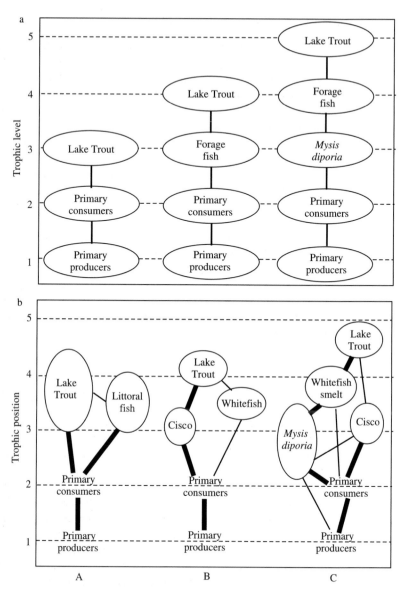

Fig. 5.18 (a) A linear food chain model of pelagic food chains in three different lake categories (A–C) differing in presence of intermediate consumers; (b) a trophic position model of the same systems where the trophic positions of organisms have been estimated using stable isotope analysis. The width of the lines linking compartments represents the importance of the interaction. From Vander Zanden *et al.* (1999).

transfers between consumers (Fig. 5.18). The trophic position model shows that organisms cannot easily be assigned to discrete trophic levels but instead have a large variability in their real trophic position due to omnivory and other complexities in the food web.

Predator identity and multiple predators

Functional groups can, as we have seen above, be a useful concept to reduce the bewildering complexity of natural food webs and allow us to study and understand the importance of complex interactions in food chains. However, species belonging to the same functional group, for example, herbivores or predators, may have different effects on the community. Large zooplankton species, for example, have higher grazing rates and affect the species composition of the phytoplankton community differently than small zooplankton species (see Chapter 4, *Biotics*) and effects on prey assemblages differs significantly when the top predator is either fish or dragonfly larvae (McPeek 1998). Thus, in order to predict the effects of changes in the abundance of one species on any other in the food web it may be necessary to have a higher resolution when we define its position in the food web. Species within a functional group, for example, predators, are often considered to be more similar in their effects on the community if they are similar with regards to different traits, including taxonomy, morphology, food resource, and habitat use. Thus, we would, for example, expect the effects among dragonfly species to be more similar than when comparing dragonfly and fish effects, effects of predators with similar gape size limitations to be similar and the effects of benthic predators to be more similar in comparison with benthic versus pelagic predators.

It is also important to remember that most experimental studies on the effects of predation on food chain interactions have involved only one predator, or in the few cases where several predator species have been studied, only 1–2 species have been manipulated at a time. This is in spite of the well-known fact that a suite of different predator species coexists in most natural communities. Recent studies have shown that the effects of multiple predator species on populations and communities cannot be predicted by simply adding the effects of individual predators; multiple predators have emergent impacts on their prey (Sih *et al.* 1998). Emergent effects of multiple predators could either result in risk-reduction or risk-enhancement for the prey. *Risk-reduction* occurs when there are interactions between the predator species that reduce predation rates for one or both of the predators. In an experiment, Fauth (1990) showed that the survival of tadpoles exposed to predation from crayfish and salamanders was lower than expected, based on single-predator treatments, in the treatment where the two predators were combined (Fig. 5.19). The reduced tadpole mortality in the combined predator treatment was due to direct physical interactions between the predators; the salamanders had injured tails and extensive scars on the body caused by crayfish. *Risk-enhancement* typically occur when the prey have predator-specific prey responses; that is, when expressing a defence adaptation against one predator results in the prey being more exposed to predation from the other predator. For example, predation rate of coexisting perch and pike feeding on roach was higher than expected based on when the two predators were alone (Fig. 5.19; Eklöv and VanKooten 2001). Perch is an open-water predator

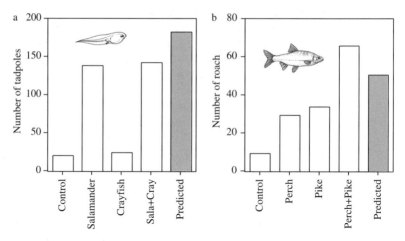

Fig. 5.19   Mortality of tadpoles and roach in experiments where the prey were exposed to predators singly or in combination. (a) Salamanders and crayfish; (b) perch and pike. The shaded bars show the predicted mortality of the combined predators based on the results from the single-predator treatments. From Fauth (1990) and Eklöv and VanKooten (2001).

whereas pike are typically ambush predators that mainly forage in or close to the vegetation. In the presence of perch alone roach moved closer to the vegetation and avoided the open-water habitat, whereas in the treatment with pike only roach used the open water almost exclusively. Thus, predator-specific changes in habitat use resulted in risk-enhancement for roach and greater than expected predation rates when the two piscivores were combined. These changes in mortality and behaviour in roach also affected macroinvertebrate and zooplankton populations.

Omnivory

Although the food web is now less complicated than the original inter-action web (Fig. 5.17), there may still exist complexities that affect the strength of indirect interactions. For example, if there is a high degree of omnivory in the system, the distinct patterns that are predicted by food chain theory should become more obscure. It has long been argued that omnivory is relatively rare in natural food webs and should thus be of minor importance in food chain dynamics. However, recent studies suggest that omnivory may be more common than was thought previously. After reviewing the relevant literature, Diehl (1993) suggested that differences in relative body size between trophic levels affect the strength of direct and indirect interactions in food chains with omnivores. When intermediate consumers and resource organisms are relatively similar in size and both considerably smaller than the top consumer, as is common in many fresh-water food chains (fish–predatory invertebrates–herbivorous invertebrates), top consumers have strong effects on lower trophic levels, whereas indirect effects are relatively weak. However, these studies did not involve primary producers, and studies including periphytic algae and crayfish have shown strong cascading effects of the omnivorous crayfish, suggesting that omnivory may not *always* result in weak indirect interactions (Lodge *et al.* 1994; Nyström *et al.* 1999).

Ontogenetic niche shifts and size-structured populations

So far, we have assumed that all individuals within a species behave in a similar way with regard to, for example, habitat choice, food preference, and predator avoidance tactics. This is, of course, a gross oversimplification. Individuals within a species do behave differently when faced with the same demands or threats. For example, quantification of diet through gut content analysis often suggests that some individuals specialize on a certain prey organism, whereas others feed on a broad range of prey items. Further, individuals within a species often experience an abrupt change in habitat preference and/or diet as they grow (i.e. they undergo **ontogenetic niche shifts**). In many fish species, larval and small juvenile fish feed on zooplankton, but then shift to include macroinvertebrates as they grow larger. In piscivorous species, the largest size-classes feed on fish, as exemplified by the diet shifts of perch (Fig. 3.22). These changes in habitat use and diet over ontogeny may have far-reaching effects on interactions among species, ultimately affecting the whole system through decoupling of consumer–resource dynamics.

Examples of ontogenetic niche shifts

In North America, Mittelbach and co-workers studied interactions between different size-stages of centrarchid fish, including bluegill and pumpkin-seed sunfish and largemouth bass (e.g. Mittelbach and Chesson 1987; Osenberg *et al.* 1992; Mittelbach and Osenberg 1993; Olson *et al.* 1995; Fig. 5.20). In the presence of the piscivorous largemouth bass, juvenile bluegills move into the littoral zone where they feed on soft-bodied benthic macroinvertebrates. Large bluegills are less vulnerable to predation by the gape-limited bass and can move out into the open water habitat and feed on zooplankton. Adult pumpkinseed sunfish are specialized snail predators, but juvenile pumpkinseed fish cannot crush snails and instead their main food source is soft-bodied invertebrates. Thus, in lakes where both sunfish species coexist with bass, the different life stages of bluegill and pumpkin-seed occupy distinctly different habitat and diet niches. Juveniles of both species exploit, and hence compete for, a common resource, the soft-bodied macroinvertebrates, whereas adult bluegills and pumpkinseeds feed on zooplankton and snails, respectively. Consequently, changes in the resource base for one of the stages may affect recruitment into the next stage, but

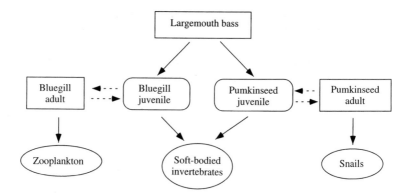

Fig. 5.20   The interactions among a piscivorous fish (largemouth bass), two sunfish with two-stage life histories (the bluegill sunfish and the pumpkinseed sunfish), and their juvenile and adult resource bases. Dashed arrows represent reproduction and recruitment between adult and juvenile stages, respectively. Based on Osenberg *et al.* (1992).

may also affect the other species through relaxed or intensified competition for the juvenile resource.

Some organisms compete at one stage of their life, but become predator and prey at another. Such an example is the fish species, largemouth bass, which is almost exclusively a piscivore when adult, but as juvenile feeds on soft-bodied invertebrates. Thus, largemouth bass are potential competitors with juvenile bluegill and pumpkinseed sunfish during their first year when they feed on invertebrates. Later in life they shift to become predators. Olson *et al.* (1995) showed in a field experiment that the growth rate of juvenile bass was significantly reduced in response to increasing densities of juvenile bluegills. Moreover, a survey of fish populations in lakes showed that the size of one-year-old bass was negatively correlated to the density of juvenile bluegills, whereas large bass had higher growth rates in lakes with high densities of bluegills. Thus, at high bluegill densities, largemouth bass suffer a cost through reduced growth rate and a delay in the shift to the piscivore stage. At the same time, however, there is an advantage for large bass individuals through a higher resource base, resulting in high growth rate and fecundity at high abundance of bluegills.

In some situations, the competition between juveniles of the predator and prey species may be sufficiently strong to severely restrict the recruitment of the predator species to the piscivorous stage. Such a *juvenile competitive bottleneck* has been shown to affect the interactions between roach and perch in eutrophic lakes (Persson 1988; Persson and Greenberg 1990). Roach (Fig. 3.21) is a cyprinid fish species with an omnivorous diet consisting of zooplankton, macroinvertebrates, and plant material. Juvenile perch feed on zooplankton and as they grow they shift to a macroinvertebrate diet and, eventually, become piscivorous. Roach are the more efficient zooplanktivores, whereas perch have a higher foraging efficiency on benthic macro-invertebrates. An increase in the density of the efficient zooplanktivorous roach should result in a decrease in the availability of zooplankton and, consequently, perch are predicted to switch earlier to feeding on macro-invertebrates. This should result in a smaller proportion of perch being recruited to the piscivorous stage; and for the few that become large enough to be piscivorous, the time spent in the benthic feeding stage should be longer than normal.

Subsidized consumers and food chain coupling

We tend to think about ponds and lakes as closed systems, and although they are more closed than most terrestrial systems, they are still open to transport of resources across habitat borders. Allochthonous material coming to the lake via inflow streams or directly from terrestrial habitats is an important energy source in many systems (see Chapter 2, *Abiotics*). In recent years, there has been a discussion on just how much of the carbon in pelagic food chains that has been fixed by phytoplankton (autochthonous carbon) and how much that has a terrestrial origin (allochthonous carbon). Autochthonous production is probably the dominant process in eutrophic lakes but in oligotrophic, and especially in humic, lakes input of allochthonous carbon is generally of higher importance. Allochthonous carbon enters the lake as particulate organic carbon (POC) and DOC, respectively). Most of the DOC that enters a lake is recalcitrant and difficult to assimilate and could thus not be readily incorporated into lake food webs. However, after degradation, for example, by UV radiation, DOC can be used by bacteria. Stable isotope studies have shown that up to 50% of the carbon in zooplankton may be derived from terrestrial sources (Grey *et al.* 2001; Pace *et al.* 2004) and, thus, carbon imported from the watershed may provide an important subsidy for lake food webs.

In modern limnology, there has also been a tendency to compartmentalize lake food webs based on within-lake habitats. Thus, organisms in the pelagic and benthic habitats are often considered to belong to discrete food chains, the pelagic and benthic food chains (Fig. 5.17), with negligible transfer of energy between them. The major focus has been on interactions and mass transfer in the pelagic food chain, but recent research has shown that carbon fixed and assimilated by benthic organisms is important for the production in lake ecosystems as the benthic and pelagic food chains are integrated by more or less complex links between them; there is a strong *benthic–pelagic coupling* (Vadeboncoeur *et al.* 2002). Studies of primary productivity has shown that benthic primary production (macrophytes and periphytic algae) may be of the same magnitude or even greater than pelagic primary production (phytoplankton). Further, analyses of the stable carbon istotope ratios ($\delta^{13}C$; reveal the origin of carbon sources) show that much of the carbon in pelagic food chains has a benthic origin. Fish is one of the strong links between the benthic and pelagic food chain and an analysis of fish diets showed that for lake fish on average 65% of its diet came directly or indirectly from benthic animals (Vander Zanden and Vadeboncoeur 2002). Thus, the energetic links between periphytic algae, macroinvertebrate grazers, detritivores, and predators and fish are important for whole-lake energy dynamics. However, the energy pathway may also go in the other direction, from the pelagic chain to organisms in the benthic habitat. For example, much of the production of benthic organisms, such as chironomid larvae and oligochaetes, in the profundal soft sediments depend on the downfall of detritus from the pelagic food chain as an energy source. Such subsidies from other food chains within the lake, or, for that matter, from the terrestrial system, may allow consumers in lake food chains to become more abundant than if they had to rely entirely on resources produced within the lake or pond and thereby affect the strength of trophic cascades (Polis and Strong 1997).

Defence adaptations against consumers

The presence of strong cascading interactions in a system requires that the majority of species within a trophic level is vulnerable to consumption, otherwise compensatory shifts among species may diffuse the cascading effects. Defence adaptations of prey tend to decouple predator–prey dynamics and decrease the strength of interactions, both direct and indirect, in food chains. Freshwater organisms have evolved a multitude of defence systems against predators, including morphological, chemical, and behavioural defences (see Chapter 4, *Biotics*).

Algae

In pelagic food chains, changes in edibility of the phytoplankton have been suggested to decrease the importance of cascading effects. As we have seen in earlier chapters (e.g. Chapter 4, *Biotics*), small, edible, and fast-growing phytoplankton species become abundant when large herbivorous zooplankton occur in low numbers (Fig. 4.16). In contrast, when such herbivores become abundant, grazer-resistant growth forms, including filaments, spines, and large cells, increase in importance. Because of these grazer-resistant morphologies, the consequences on the total algal biomass of a shift from small, inefficient, to large, efficient, zooplankton, may not always be as drastic as expected. Instead, when an algal community changes from being dominated by species vulnerable to grazing to one dominated by grazer-resistant species, the total algal biomass may, in some cases, be the same! The shift to inedible forms is one of the major reasons why food web theory sometimes fails to predict what we actually see in natural systems.

Grazer-resistant bacteria

As we concluded when discussing the microbial loop, bacteria not only suffer from a very high grazing pressure mainly from heterotrophic flagellates and ciliates, but also from large macrozooplankton such as *Daphnia*, which calls for grazer-resistant adaptations. Since bacteria probably were among the first organisms on our planet, they have had plenty of time to develop adaptations against grazing. Moreover, their generation time is very rapid, suggesting that sophisticated adaptations may have had time to evolve. This is, surprisingly, not the case and bacteria have, compared to more complex organisms, relatively inefficient methods of avoiding grazers. Just as for other prey organisms, one way for bacteria to avoid being consumed is to be smaller or larger than the optimal size requirement of their grazers—heterotrophic flagellates and ciliates. Although the last two organisms cannot be said to be 'gape-limited' like fish predators, very small and very large bacteria are outside the optimal *prey size window* of the grazer, thereby reducing feeding efficiency. With respect to bacteria this includes growth forms such as very *small spheres* ($<0.05\ \mu m^3$), or bacteria that become enlarged by forming *filaments*, *spirals*, and *aggregates*. Based on several laboratory experiments, Jürgens and Güde (1994) attempted to generalize a typical succession pattern of bacteria and their predators (heterotrophic flagellates and ciliates; Fig. 5.21). They separated the succession into three phases: In phase I, rod-shaped bacteria increase in biomass. In phase II, heterotrophic flagellates develop and reduce the biomass of rod-shaped bacteria to very low levels. Simultaneously, with a high biomass of flagellates the bacterial biomass again increases, but now as a morphologically more complex community, including spirals, filaments, and aggregates.

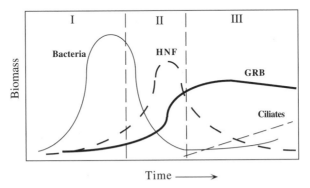

Fig. 5.21   Idealized microbial succession divided into three phases. Phase I is characterized by high bacterial growth rate and a subsequent increase in HNF feeding on the bacteria. In phase II, the bacterial abundance declines and HNF reach their highest biomass which eventually decreases due to low food supply. Moreover, GRB, including filaments and spirals, increase in importance. In phase III, ciliates, which are large enough to feed on the GRB, increase in importance. Modified from Jürgens and Güde (1994).

These are probably less edible to flagellates, which decrease in biomass (phase III, Fig. 5.21). Later during this phase, ciliates appear that probably are more able than the flagellates to eat the grazer-resistant bacteria (GRB). In this way, the total bacterial biomass is more or less constant over time, but comprises different growth forms. Although filamentous and aggregated bacteria may have an advantage at high grazing pressure, they are most probably less competitive with respect to substrate than spheres and rod-shaped cells. Hence, as for most other organisms there is a trade-off between getting enough food and reducing the risk of being eaten!

Gape-limited predatory fish

Most fish consume their prey whole and thus the largest prey that these predators can ingest is limited by the size of their gape; these are *gape-limited predators*. Consequently, prey that are larger than the largest gape of the predator have reached an absolute size-refuge and are no longer susceptible to predation. Hambright *et al.* (1991) suggested that absolute size-refugia may have important implications for interactions in freshwater

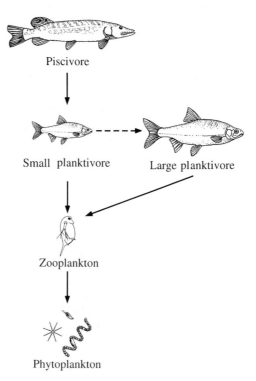

Fig. 5.22   Decoupling of cascading trophic interactions in a pelagic, freshwater food chain. Large planktivores have reached a size-refuge from predation by the gape-limited piscivore and thus can have a significant effect down the food chain, even in the presence of piscivores. The broken line denotes recruitment of large planktivores from the small planktivore pool.

food chains. In some systems, predation by gape-limited piscivores may result in prey fish populations dominated by large, deep-bodied planktivores that have reached an absolute size-refuge (Hambright *et al.* 1991; Hambright 1994). Predation by these large planktivores on zooplankton may still be so intense that they reduce the density of large, efficient cladocerans and thereby affect phytoplankton biomass. The cascading effects from piscivores are thus decoupled at the piscivore–planktivore level (Fig. 5.22) due to an absolute size-refuge in the planktivores.

Structural complexity

High structural complexity may also affect the strength of the predator–prey interaction by reducing the predator's foraging efficiency and by providing refugia for prey organisms and thereby decreasing the effect of indirect interactions. With respect to pelagic bacteria, the only spatial refuges are suspended particles, which often harbour large amounts of bacteria. Although these particles may be refuges against grazing from protozoan grazers, they do not provide any protection against grazing by *Daphnia*. At the sediment surface, however, large particles are efficient refuges against grazers, which together with a high carbon and nutrient availability, result in very high bacterial abundance at the sediment surface compared to those in the water.

With respect to larger organisms, the habitat with the highest complexity is generally the littoral zone with submerged and emergent vegetation. A number of descriptive studies of invertebrate populations inhabiting macrophyte beds have shown wide diversity and biomass of benthic macroinvertebrates (e.g. Soszka 1975; Dvorac and Best 1982). Diehl (1988) showed that the density of chironomids was higher in microhabitats with submerged vegetation (Fig. 5.23(a)) and this was related to a lower foraging efficiency of benthivorous fish such as perch (Fig. 5.23(b)) in the more complex habitat.

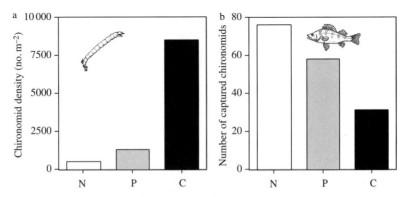

Fig. 5.23 (a) The density of chironomids in habitats with different submerged vegetation; (b) the foraging rate of perch feeding on chironomids in habitats with different submerged vegetation. N, no vegetation; P, *Potamogeton*; C, *Chara*. After Diehl (1988).

Succession

Seasonal succession among organisms is one of the more obvious and regular features of lakes and ponds. Hence, some organisms always occur during spring, whereas others exclusively dominate during summer or in late autumn, forming a succession pattern that is generally repeated every year. Several factors are responsible for this repetition. Some organisms are dominant during certain periods of their *life cycle* and then disappear, such as chironomid larvae occurring in several thousand individuals per square metre of sediment surface. Then, suddenly, they emerge and leave the aquatic environment to become terrestrial midges. Other factors affecting succession are different *physical or chemical requirements* of organisms, including temperature and nutrient or food preferences. Interactions among the organisms themselves are also important, including *predation, grazing,* and *competition* pressure. Hence, following a seasonal cycle in a lake will reveal that abiotic as well as biotic processes act to shape the observed succession pattern among organisms. Succession patterns are therefore elegant illustrations of how the abiotic frame and biotic interactions together shape what we see when we investigate a lake or pond.

Factors behind succession patterns

How abiotic factors, such as temperature and the supply of nutrients, together with different competitive abilities may create a succession pattern among algal species can be illustrated with a laboratory experiment carried out on algal communities from three North American lakes (Lake Superior, Lake Michigan, and Eau Galle Reservoir; Tilman *et al.* 1986). The laboratory cultures were supplied with different ratios of silicon (Si) and phosphorus (P) at different temperatures (9° and 15°C). At 9°C, and an Si : P ratio close to 1, green algae and diatoms constituted about 50% each of the total number of algae. At higher Si : P ratios, diatoms became dominant, whereas green algae decreased in abundance. At higher temperatures, however, green algae were completely dominant up to an Si : P supply ratio of about 5, where diatoms increased in number and had taken over the dominance at an Si : P ratio of about 70. This experiment nicely illustrates the point that dominance in algal communities is affected by nutrient competition, but that the outcome of competition may be affected by abiotic factors such as temperature. Thus, we have identified some important forces behind seasonal succession in organism communities: not one variable or process alone, but several variables interact to shape succession and dominance patterns in aquatic ecosystems.

One of the more thorough attempts to describe systematically and explain seasonal succession in the planktonic community of lakes is the

PEG (Plankton Ecology Group) model (Sommer *et al.* 1986). The original model consists of 24 statements based on general patterns found in lakes and ponds. Below, we briefly go through some of the statements which cover the seasonal succession in an idealized lake from late winter, through spring, summer, and autumn to the next winter.

The PEG model

WINTER

A. Towards the end of winter, nutrient availability and increased light permit unlimited growth of phytoplankton. Small, fast-growing algae, such as diatoms, dominate.
B. The small algae are grazed by herbivorous zooplankton, which become dominant.

SPRING

C. Grazing rate exceeds phytoplankton production rate, leading to a reduction in algal biomass. This period of relatively clear water is referred to as the **clear water phase**. Nutrients are recycled by the grazing process and accumulated in the water.

SUMMER

D. Herbivorous zooplankton become food-limited and fish predation increases (this is due to higher temperature and the development of young-of-the-year fish), leading to a reduction in size and abundance of zooplankton.
E. As a result of high nutrient availability and low grazing pressure, a diverse summer phytoplankton community starts to build-up, including large inedible colonial green algae and small edible *Cryptophytes* (e.g. *Cryptomonas*, Fig. 3.6).
F. Algal growth becomes phosphorus-limited and the colonial green algae are replaced by large diatoms.
G. Silica depletion leads to replacement of the large diatoms by large dinoflagellates and/or cyanobacteria.
H. Nitrogen depletion favours a shift to nitrogen-fixing species of filamentous cyanobacteria.
I. Small crustacean zooplankton and rotifers, which are less sensitive to fish predation, become dominant in the zooplankton community.

AUTUMN

J. Physical and chemical changes, such as mixing followed by nutrient replenishment, lead to large unicellular or filamentous algae, and eventually diatoms become dominant.
K. This assemblage is accompanied by small edible algae, which together with reduced fish predation (lower temperature), lead to an autumnal maximum of zooplankton including larger forms.

WINTER

    L.   The reduced temperature and light energy input lead to lower primary production and a decline in algal biomass.

    M. The fecundity of many zooplankton species is reduced and some produce resting stages. However, some cyclopoid copepods contribute to the overwintering zooplankton assemblage.

Relative strength of factors behind succession patterns

Fluctuations in biomass of phyto- and zooplankton, and the relative strength of physical (mainly light and temperature), chemical (mainly competition for nutrients) factors, and predation/grazing can also be illustrated graphically as in Fig. 5.24. We see that the relative strength of physical factors is highest during autumn and winter. Moreover, biotic interactions, such as predation and grazing, are mainly important during spring to autumn. We also see that the pronounced spring bloom of algae is followed by a short maximum of zooplankton (the 'clear water phase') in eutrophic lakes. Although most temperate lakes actually follow the pattern described by the PEG model, there are many exceptions from this idealized view. However, despite the fact that nature exhibits more variation than the human brain can accommodate, attempts to structure and find general patterns are extremely important for our understanding of how ecosystems function.

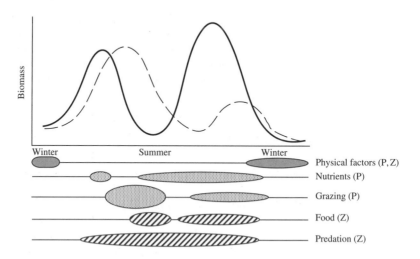

Fig. 5.24   Idealized seasonal development of phytoplankton (top panel; solid line) and zooplankton (broken line) in a stratified eutrophic lake. The elliptic diagrams show when major abiotic factors, such as temperature or nutrient limitation, and biotic interactions are important for phytoplankton (P) and zooplankton (Z), respectively. Based on Sommer et al. (1986).

Impact of organisms on the abiotic frame

Historically, a relatively static view has been applied to aquatic ecosystems implicitly stating that abiotic factors (i.e. the chemical and physical environment) constitute the frame to which organisms have to adapt. This static view is true, in part: organisms trying to invade a polar lake will, for example, have to get used to the low temperature and the long dark winter. Similarly, occasional dry-outs are nothing the organisms spending their lives in a small pond can do anything about. However, the activities of organisms can in many respects *affect and shape the abiotic frame*. Imagine a lake or pond without any organisms, as was the case before life entered Earth. In such a situation, chemical and physical processes will go on without being affected by any biological influence. If we add primary producers, for example, say algae, changes start to take place. First, the oxygen level increases due to photosynthesis, which affects oxidation–reduction reactions between chemical components. If the algae happen to be attached to the sediment surface, their photosynthesis will oxidize the surface sediment, thereby reducing the phosphorus outflow from sediment to water, since oxidized iron (Fe^{3+}) will bind to phosphate and the complex will stay in the sediment (see Chapter 2, *The abiotic frame*). The primary producers may also increase the pH, which is a result of carbon dioxide (CO_2) being taken from the water to fuel the photosynthesis. Since the photosynthesis is switched on only during daytime, pH in the lake water will show diurnal oscillations, especially in eutrophic lakes where phytoplankton photosynthesis is intense. Similarly, pH is generally higher during summer than during winter in temperate lakes. In this way, the primary producers indirectly affect the conditions of life for other organisms by creating alterations and oscillations in the chemical environment. Below we provide some surprising examples of how organisms affect the abiotic frame.

Effects of organisms on the water temperature

Adding higher organisms to our imaginary lake reveals even more sophisticated examples of how organisms can affect the abiotic frame. One striking and surprising example is that high fish abundance may affect the physical environment by reducing the water temperature. The logic behind this statement is that fish eat zooplankton, which are capable of reducing the phytoplankton biomass. At high fish abundance, the water therefore turns green and the sunlight does not penetrate as deeply as in clear lakes. This means that a larger portion of the solar radiation is scattered back into the atmosphere by the phytoplankton particles and, accordingly, that a smaller portion of the heat is absorbed by the water in turbid than in clear lakes. The obvious result of this is that the **thermocline** (i.e. the mixing depth) becomes more shallow in lakes with fish than without fish. This was shown convincingly in some Canadian lakes

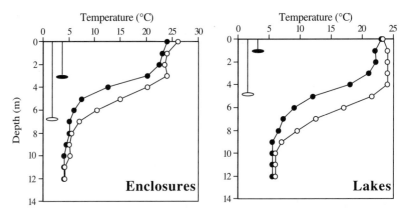

Fig. 5.25   Temperature profiles and water transparency (Secchi depth, vertical lines) from enclosures with fish (closed symbols), and without fish (open symbols), and from lakes with high (closed symbols) and low (open symbols) abundance of planktivorous fish. Data from Mazumder *et al.* (1990).

(Mazumder *et al.* 1990). In lakes and experimental enclosures with high fish abundance, the water transparency (Secchi depth), temperature, and heat contents were lower than at low fish abundance (Fig. 5.25). Differences were most pronounced in July and August when temperature at the same depth could be more than 5°C lower and the thermocline 2 m shallower in lakes and enclosures with high compared to low fish abundances.

Organism interactions affecting carbon input from the atmosphere

In a series of lakes it was shown that at high abundance of planktivorous fish, the zooplankton community became dominated by small organisms, and phytoplankton primary production increased compared to that in reference lakes with a lower abundance of planktivorous fish (Schindler *et al.* 1997). This hardly surprising result was in line with food web theory. However, the researchers were also able to demonstrate that the increase in primary production increased the input rate of carbon from the atmosphere to the lake (i.e. the increased 'speed' of the primary production 'pulled' more carbon into the lake). The atmospheric carbon was then traced in algae, as well as in zooplankton and fish (Schindler *et al.* 1997). Hence, in this example, the abundance of fish affected a fundamental biogeochemical process that had far-reaching consequences for many ecosystem processes.

With these examples, we have shown that the abiotic frame is not static, but may be affected by the inhabitants of a lake or pond. By applying the knowledge from this book, in combination with your own imagination, you may easily come up with additional examples of how organisms in a specific lake can affect the abiotic frame. For example, what are the consequences when macrophytes transport oxygen from the atmosphere to their roots, and subsequently leak oxygen to the sediment? Or, how is the nutrient concentration in the water column affected by massive recruitment of cyanobacteria from the sediment, absorbing nutrients at the sediment surface, and then migrating and eventually dying in the open water?

However, we will now leave idealized natural conditions and devote the next Chapter on an organism that during recent time has proven highly efficient in affecting both the abiotic frame and the biotic interactions in freshwater ecosystems: *Homo sapiens*.

Practical experiments and observations

Non-lethal effects in food chains

Background The mere presence of a predator may change the behaviour (e.g. activity level) of a herbivore or detritivore and this may then affect the accumulation of primary producers and detritus, respectively (e.g. Turner 1997). In this experiment you can investigate the effect of fish cues on the decomposition rate of leaf detritus. Fill a series of containers with lake water and add several alder leaves that have been dried and weighed. Add a shredder (e.g. caddisfly, water louse (*Asellus*), or freshwater shrimp, *Gammarus*), to treatment containers, but keep some containers as shredder-free controls. During the experiment add water from tanks with fish once a day to half of the treatment containers. Let the experiment run until the shredder has had a noticeable effect on the leaves in at least some containers. Remove the leaves, dry them, re-weigh them, and calculate losses due to shredder activity. In addition, you could monitor the activity levels (e.g. number of moving individuals) of the shredders in the different treatment containers. The experiment could be repeated with shredders that have different defence adaptations (e.g. cased caddisflies versus freshwater shrimps). Further, the nature of the fish cue can be altered by using different fish species, comparing starved and fed fish, and by comparing fish that have been fed different diets.

To discuss Is there any effect of treatments on leaf litter breakdown? What is the reason for the differences? What conclusion can you draw from a small-scale experiment like this?

Pelagic food chains

Background Fish predation may have a strong impact down pelagic food chains, affecting phytoplankton through a reduction in the density, and hence grazing pressure, of large, daphnid zooplankton. This has been shown in experiments at different scales, but most commonly in large-scale field exclosures. The potential for strong cascading effects could also be shown, however, in small-scale laboratory experiments.

Performance Set-up a series of containers (e.g. aquaria or even large buckets) and fill them with lake/pond water that has been filtered to remove zooplankton. Add a few drops of commercial plant nutrient to each container to enable algae growth. Divide the containers into three groups (treatments) and let the first group remain as it is (i.e. have only algae; one trophic level). Collect large daphnids (easiest in a pond without fish) and add them to the second group of containers (density: about $30\,l^{-1}$) which then has both algae and zooplankton (two trophic levels). To the third group of containers, which should have three trophic levels, add zooplankton as well as one or two small fish (sticklebacks, guppy). Quantify the zooplankton density and phytoplankton biomass when the experiment is terminated after two weeks. Differences in dominant zooplankton and phytoplankton species could also be studied. The experiment could be done either with a low or a high input of nutrients to determine if top-down effects are equally strong at different nutrient loads.

To discuss In which treatment is the phytoplankton biomass highest? What affects the algal biomass? Are top-down or bottom-up effects most important?

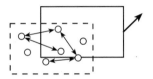

6 Biodiversity and environmental threats

Introduction

The water in lakes and ponds is only a minute fraction of the total water resource of the earth. A large proportion of the freshwater resource is stored as ice and snow at high altitudes and around the poles (about 77%) or below the ground (22%); less than 0.5% is available for use by organisms. Water availability is a cornerstone for human civilization, but increasing human populations have resulted in accelerating demands on water supplies for drinking, hygiene, industrial processes, and agriculture (irriga-tion). Analyses of water availability and human use suggest that human withdrawal of the total available freshwater resource presently amounts to approximately 50% (Szöllosi-Nagy *et al.* 1998). The expected population increase coupled with economic development and changing lifestyles in the future will substantially increase the demand for freshwater resources, and without doubt the availability of freshwater for human consumption will be one of the great issues for humankind in the twenty-first century (Johnson *et al.* 2001). Further, organisms living in freshwater systems, such as fish, water plants, and invertebrates, are exploited by humans for food, and in many parts of the world food products with freshwater origin make-up an essential part of the diet.

Even though the access of freshwater of high quality is of the greatest importance to us, we have misused this resource for many centuries. Most freshwater systems have been heavily affected and disturbed by our activities, including eutrophication, chemical pollutants (herbicides, insect-icides, and heavy metals), acidification, drainage systems, irrigation systems, landfilling, and introduction of exotic aquatic species. This not only affects the use of freshwater systems as a resource for humans, but also the integrity of freshwater systems as biological communities. Freshwater systems harbour a unique and diverse set of species. About 15% of all animal species that exist today live in different freshwater systems. More than 70 000 freshwater species from 570 families and 16 phyla have been described so far (Strayer 2001), but a similar number may remain to be discovered and described as the knowledge of taxonomy and distribution of freshwater organisms are poor in many parts of the world. In lakes and ponds, most species belong to crustaceans, rotifers, insects, or oligochaetes. Planktonic and periphytic algae as well as different life forms of macrophytes are also speciose. However, species and even whole communities are now being lost at an alarming rate. Extinction rates in freshwater systems may be as high as for tropical rainforests (Ricciardi and Rasmussen 1999), which are considered to be the most stressed terrestrial systems on earth. In fact, recent data suggest that freshwater biodiversity has been reduced at a higher rate than marine or terrestrial biodiversity over the past 30 years (Jenkins 2003). Presently, more than 1100 freshwater invertebrates are endangered, a number that is most certainly too low as knowledge on smaller, less conspicuous, or economically unimportant species is sparse and there is little or no monitoring of freshwater organisms in large parts of the world (Strayer 2001). Some groups seem to be more at risk than others. For example, 21 of the 297 North American freshwater clam and mussel species are already extinct and over 120 are threatened, and 30% of North American and 40% of European fish species are threatened. There is also a global decline in amphibian populations (e.g. Houlahan *et al.* 2000).

Many factors affect biodiversity in natural systems. Changes in land use, atmospheric CO_2 concentration, nitrogen deposition, acid rain, climate change, and species invasions are the most important threats to biodiversity at the global scale (Sala *et al.* 2000). For lakes, land use changes and invasion of exotic species will remain major drivers of changes in biodiversity over the next century. Agricultural non-point pollution resulting in increased nutrient loading and high concentrations of toxic contaminants, exotic species, and habitat fragmentation are the three major threats to the freshwater fauna (Richter *et al.* 1997). Further, several factors often act in concert to cause the extinction of a certain species. For example, the number of threats affecting endangered freshwater species in the United States ranges from one to five per species (average: 4.5; Richter *et al.* 1997). Before we go into further detail on the different threats to lake and pond systems we will focus a bit more on biodiversity, that is, what biodiversity really is, what determines biodiversity at different spatial scales and if there are any patterns in biodiversity. This is because biodiversity may be regarded as the ultimate response variable measuring the effect of disturbances on the natural system. Then we will look into the most important environmental threats to lake and pond systems and if it is possible to *restore*, or at least *rehabilitate* degraded lakes and ponds.

Biodiversity in lakes and ponds

Definition

Biological diversity, or biodiversity, has in recent years become a buzz-word used in many different situations and although most of us have an intuitive feeling for what the word stands for we may still need to define it. *Biodiversity* could be broadly defined as *the variety and variability among living organisms and the ecological complexes in which they occur.* Biodiversity could be considered at different scales ranging from the gene to the ecosystem. *Genetic diversity* refers to the variation in gene frequencies within and among populations of the same species, for example, among populations of fish from different lakes and watersheds. Preserving genetic diversity can be an important issue when a species shows high variation in, for example, morphological or life history characteristics among populations from different habitats due to adaptation to the local environment. Loss of genetic diversity affects a species ability to adapt to new diseases or changes in environmental conditions, natural or human-induced. *Organismal diversity* refers to the number of species living in an area, such as a pond or a lake, whereas *ecosystem diversity* is the diversity of habitats and ecosystems within a region. The most commonly used meaning of biodiversity is at the level of species (organismal biodiversity) and that is also the level we will focus on here. However, it should be noted that many actions by man changes biodiversity of freshwaters at the ecosystem level as well. Drainage, landfilling, and other types of habitat destructions clearly decrease the availability of natural freshwater habitats in a region and thereby also affect ecosystem biodiversity. On the other hand, recent restoration efforts may increase ecosystem biodiversity, for example, the building or restoration of wetlands and ponds to increase nutrient retention in watersheds.

Measures of biodiversity

In order to study, for example, how organismal biodiversity differs between systems or how biodiversity is affected by human activities such as eutrophication, we need a way to measure biodiversity. *Species richness*, that is, the number of species in the environment studied, is the simplest and commonest measure of biodiversity. However, a problem with just counting the number of species when estimating biodiversity is that this measure does not account for differences in abundance among species, that is, if some species are very common and some rare in the system. To illustrate this we may imagine two hypothetical pond systems, each containing 100 individuals of 5 different fish species. In the first pond there is a relatively even distribution of individuals among the five species, whereas in the second

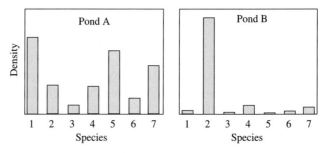

Fig. 6.1   The density of fish in two ponds. Both ponds have the same species richness (seven species) but differ in diversity as calculated by the Shannon–Weaver index.

pond one species is dominant and the other occur only at low densities (Fig. 6.1). Intuitively, the diversity of the first pond is higher than in the second, which has almost a monoculture of species 2. Species richness, however, is identical for the two ponds and, thus, we need some other measure of biodiversity that takes relative abundances of the species into account. Two common diversity indices that do exactly that are Simpson's index (D) and the Shannon–Weaver index (H). Both use the proportions (p_i) of the species (i) of the total number of organisms in the assemblage (i.e. the number of individuals of species i divided by the total number of individuals in the assemblage). Thus,

$$D = 1 / \Sigma \ (p_i)^2$$

and

$$H = - \Sigma \ p_i \ln p_i$$

If we instead use these indices to calculate the diversity of the two fish ponds we get $H = 1.77$ for the first pond and $H = 0.76$ for the second suggesting differences in diversity even though species richness is identical.

Using these indices may be a better measure of biodiversity than just species richness, but we may still find situations where this is not enough. Imagine two lakes, each inhabited by five fish species. The first lake has a species assemblage that consists of five different planktivorous fishes, whereas the second lake has two planktivorous fish species, a benthivorous species, and two piscivorous fish, and they occur at a similar frequency distribution as in the first lake. Now, even though diversity, as measured by either species richness or the Shannon–Weaver index, is identical between lakes, a fisherman would immediately tell you that the diversity is highest in the second lake. This is because we intuitively think of diversity as the variability among organisms in a community and some aspects of this variation are not captured by simple indices. In this case we perceive a higher diversity in the second lake because it has fish that feed on different trophic levels and in different habitats as compared to the first lake where all five fish species feed on plankton in the pelagic zone. Hence, we have a higher functional diversity in the second lake, that is, a larger range of different species traits in the species pool. When considering the rate and efficiency of ecosystem processes a higher functional diversity may be more important than just a higher number of species (see Biodiversity and ecosystem function below).

What determines biodiversity in ponds and lakes?

The species richness of a specific pond or lake is a function of a number of processes operating at different spatial and temporal scales, ranging from historical and global/regional factors such as colonization and extinction to present-day abiotic and biotic interactions in the local environment. A useful approach to understand the interaction of processes at the different scales is to view the species composition in a specific lake as the result of a series of different filters that operate at the regional to the local scale (Tonn *et al.* 1990; Rahel 2002; Fig. 6.2). Factors operating at the local scale, for example, competition and predation, can only affect the species that have colonized from the regional species pool. The size and composition of the regional species pool, that is, the number of species available for colonization of local lakes in a region, is determined by factors operating at the larger geographic and historical scale. At the largest, biogeographic scale we can consider plate tectonics, that is, the movement and collisions of continents, that may have determined why some taxonomic groups are present or absent on the continental scale. Within a continent, factors such as speciation and extinction,

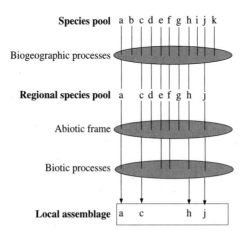

Fig. 6.2   A conceptual model showing the factors determining which species are present in the local assemblage of a specific lake as a result of 'filters' operating at different spatial scales. Modified from Tonn *et al.* (1990) and Rahel (2002).

climatic factors, range extensions and contractions, glaciation and dispersal barriers affect the number of species available in the regional species pool. A region may be defined as a geographic area where there are no significant barriers to dispersal and for most freshwater organisms a region could be conveniently defined as a catchment. Freshwater organisms use either active or passive dispersal mechanisms when colonizing new habitats (Bilton *et al.* 2001). *Active dispersal* is when individual organisms move by themselves between waterbodies. For example, freshwater insects with an adult, winged stage can fly to and oviposit in a different waterbody than the one where they spent their larval period. *Passive dispersal* occurs when the organism is dependent on some external transport mechanism for dispersal. A number of more or less anecdotal evidences of passive dispersal in freshwater organisms have been recorded over the years (Bilton *et al.* 2002). Snails, mussels, amphipods, etc. have been found attached to waterfowl, amphibians, and insects. However, the importance of dispersal by such animal vectors is not known. For smaller organisms like zooplankton, dispersal by wind may also be important (Cohen and Shurin 2003).

On a smaller spatial scale, biodiversity is determined by both abiotic and biotic factors. The abiotic frame provided by the local lake determines which species are available for colonization and which are filtered out (Fig. 6.2). Abiotic factors include, for example, temperature regime, pH, nutrient availability, habitat stability, and complexity. Most species have a rather narrow range of environmental conditions within which they exist. If, for example, the winter or summer temperature is very low or high, respectively, or if pH levels decrease below a threshold value many organisms will be unable to grow, reproduce, and eventually survive in the habitat. Structural complexity is another factor affecting species diversity by increasing number of available niches in the habitat. Macrophytes, both submerged and emergent, increase the structural complexity of lakes and ponds which results in an increase in the diversity of attached algae (epiphyton), invertebrates, fish, and waterfowl by providing structure to grow on, a food resource and a refuge from predation.

Not all species that fit into the abiotic frame of a lake are able to colonize and persist. Ponds and lakes could be seen as biogeographic islands isolated in the terrestrial landscape and thus the well-known island biogeography theory developed by MacArthur and Wilson (1967) may be applied to these systems. According to this theory the equilibrium number of species on an island is dependent on immigration and extinction processes. Immigration of new species from a species pool to an island, or in this case a lake, will decrease as the number of species already present on the island increases. Extinction rate, on the other hand, will increase as the number of species on the island increases. Larger islands may be easier to find for a colonizing organism resulting in larger islands having higher number of species. This may also result from decreasing extinction rates with increasing island area as the smaller populations on small islands are more prone to extinction than the larger populations that can exist on the larger islands. Thus, the number of species on an island is predicted to increase with island area. A number of studies have shown that this relationship also holds for freshwater organisms colonizing lakes including zooplankton, snails, and fish (Browne 1981; Fig. 6.3).

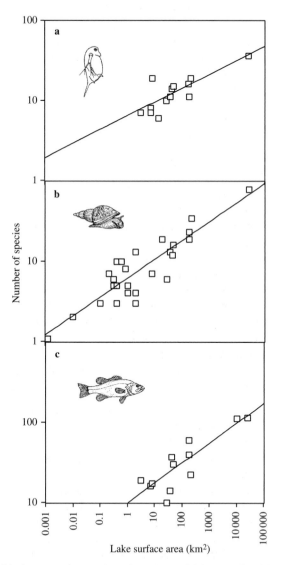

Fig. 6.3   The relationship between the number of species and lake area for: (a) zooplankton; (b) snails; and (c) fish in North American lakes. From Browne (1981).

Biotic processes are important at the local scale in freshwater systems. A number of studies have shown that predation and competition may determine species diversity and, maybe more importantly, species composition, and thereby the function and dynamics of the whole system (see Chapters *Biotics* and *Food web interactions*). Moderate levels of grazing or predation may actually increase species diversity of prey organisms. Grazing and predation are examples of disturbances that may halt the succession and promote species diversity. A general hypothesis explaining the effects of disturbances, including also physical disturbance events, was proposed by Connell (1978). The *intermediate disturbance hypothesis* suggests that diversity shows a hump-shaped relation with disturbance with highest diversities at intermediate disturbance levels. At low levels of disturbance population densities will be high, competition for resources intense, and competitive exclusion will reduce diversity to low levels as only the dominant competitors will survive. At high disturbance levels, on the other hand, only those species that are resistant to the disturbance or that colonize quickly can subsist, resulting in low diversity. At intermediate frequencies or intensities of disturbance, population densities of the competitively dominant species will be reduced and resources will be released for use by less competitive species enabling them to coexist and resulting in high levels of diversity. Intermediate disturbances have been suggested as one explanation of the 'paradox of the plankton'. Experiments on phytoplankton cultures where the frequency of disturbances varied resulted in highest species diversity at intermediate disturbance frequencies (Flöder and Sommer 1999; Fig. 6.4).

Thus, local diversity is a function of processes within the lake and 'filters' at higher levels that determine which species could enter the lake. Anthropogenic disturbance, that is, the environmental threats that we will deal with in this chapter, affects the position of the abiotic frame resulting in that some species may find themselves outside the frame; they will be

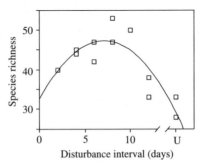

Fig. 6.4   The species richness of phytoplankton in experimental enclosures at the end of the experiment. The enclosures were disturbed at different intervals by breaking the thermocline with compressed air bubbles. U, undisturbed. From Flöder and Sommer (1999).

filtered out (Fig. 6.2). One should also note that many freshwater organisms, such as amphibians and many insects, have a complex life cycle where they spend some stages of the life cycle (adults) in terrestrial environments. Factors affecting the growth and survival of the terrestrial stage will thus affect the aquatic stage and, therefore, interactions in the terrestrial surroundings may have important repercussions on freshwater biodiversity. Further, human introductions of species to areas outside their biogeographic range circumvent the biogeographic filter and if they can pass through the abiotic/physiological and biotic filters they become successful invaders of the local community (see Exotic species below).

Patterns in biodiversity

Latitudinal gradients

We know from many habitats that there are patterns in biodiversity at different scales, from global to local. On the largest, global geographical scale it is well known that number of species is higher in the tropics than in the temperate or Arctic zone. For example, there is a total number of 725 freshwater fish species in Venezuela but only 49 in Sweden. There are many different explanations for this pattern abound in the literature, including differences in energy input, climatic variability, spatial heterogeneity, disturbances, importance of competition, etc. (see, for example, Krebs 2001). The causes for the gradient in biodiversity from the poles to the tropics may vary among species and over evolutionary and ecological time different factors have certainly varied in importance and also interacted in complex ways. In freshwater organisms, however, the global patterns of biodiversity are less clear than in other species and in some groups latitudinal trends may even be opposite to the common patterns. An analysis of changes in diversity along the latitudinal gradient for organisms from different taxa revealed that body size affected the strength of the latitudinal gradient (Hillebrand and Azovsky 2001). Small organisms, such as meiofauna, protozoa, and diatoms showed weak or no correlations to latitude whereas larger organisms (trees, vertebrates) had a strong negative correlation to latitude. It was suggested that a high dispersal ability of small organisms results in global distributions and, thus, weak biogeographic distribution patterns.

Biodiversity versus productivity

Many studies have shown that species richness changes with increasing productivity of the habitat, but the exact shape of this relationship and the mechanisms that determine this has been the subject of a long-lasting controversy in ecology. Knowing the shape of this relationship may be fundamental for our understanding of how anthropogenic increases in productivity, for example, eutrophication of lake systems, affect biodiversity

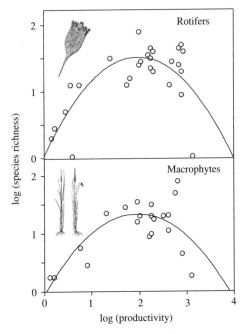

Fig. 6.5   The relationship between species richness and productivity for rotifers and macrophytes. From Dodson *et al.* (2000).

and thereby the potential to manage, preserve, or restore natural systems. Some studies suggest a positive, linear increase in species richness, whereas most empirical studies in aquatic systems suggest a unimodal ('hump-shaped') relationship between species richness and productivity with highest species richness at intermediate productivity levels (Dodson *et al.* 2000, Mittelbach *et al.* 2001). Analysis of annual primary productivity and species richness showed a unimodal relationship for phytoplankton, macrophytes, copepods, rotifers, cladocerans, and fish (Fig. 6.5). The peak in species richness differed between taxa but generally occurred at productivity levels typical for oligotrophic to mesotrophic lakes, whereas species richness generally decreased as lakes became eutrophic. Several mechanisms may explain the hump-shaped relationship between species richness and productivity. At the low end of the richness–productivity gradient an increase is expected as there is a higher potential to support additional species and trophic levels as available energy in the system increases (Oksanen *et al.* 1981, see Fig. 5.5). At the high end, several mechanisms have been proposed to explain the decrease in species richness with primary productivity, including changes in competition intensity, predation pressure, and the importance of abiotic factors with increasing productivity. For lakes, the fact that only few organisms can tolerate the extreme abiotic conditions at the upper end of the productivity gradient, that is, hypertrophic lakes, with low oxygen concentrations, high pH, and low light availability, may be the cause for the decline in species richness (Dodson *et al.* 2000). In addition, human activities that result in eutrophication of lakes to hypertrophic conditions may also be accompanied by contamination by other toxicants also affecting species richness. Even though the exact mechanisms creating the patterns remain to be found we have here seen that species richness can be a function of productivity. Below, we will look at the other side of the coin, and discuss how productivity, and other ecosystem processes, instead may be a function of biodiversity.

Biodiversity and ecosystem function

Recently, there has been an increasing concern about the ecological consequences of changes in biodiversity as the number and composition of species also determine which organismal traits are present and can influence ecosystem processes. Already Darwin suggested that diversity could affect productivity of ecosystems. He noticed that a plot sown with several distinct species of grasses gave a higher yield than a plot sown with only one species of grass. But even if it is not a new idea, the accelerating loss of global biodiversity has intensified the search for general relationships between species diversity and processes that determine ecosystem functioning. *Ecosystem functioning* refers to the biogeochemical processes within the ecosystem, that is, processes that contribute to the transformation of energy and matter in the system, including processes such as primary productivity, consumption, decomposition, respiration, and nutrient uptake and retention. Further, stability properties of the ecosystem, such as resilience and resistance to external perturbations, are commonly related to species diversity.

A number of simple, hypothetical relationships between species diversity and ecosystem functions have been suggested based on both theoretical analyses and empirical studies (e.g. Lawton 1994, Loreau *et al.* 2002; Fig. 6.6). The *null hypothesis* is of course that that there is no change in

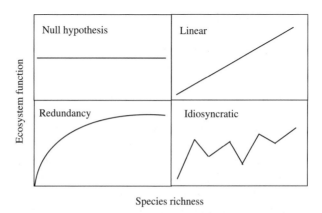

Species richness

Fig. 6.6   Hypothetical relationships between ecosystem function and species richness.

the ecosystem variable under study with a loss of species. However, the traditional argument has been that there is a *linear, positive relationship* between species richness and ecosystem processes (MacArthur 1955). A linear relationship implies that species contribute to ecosystem processes in ways that are unique and a loss of a species results in a detectable change in ecosystem functioning. The *redundant species hypothesis* suggests that many species are redundant in an ecosystem and ecosystem function will not change as long as all functional groups are represented (Walker 1992). Thus, along the major part of the relationship there will be no or very little change in ecosystem function with a loss of species. In contrast, in systems where the functional roles of individual species are complex and varied the magnitude and direction of change in ecosystem function with loss of species will be context-dependent and unpredictable; the *idiosyncratic response hypothesis* (Lawton 1994).

Two different mechanisms have been suggested as the explanation to an increase in ecosystem functioning with increasing biodiversity (e.g. Kinzig *et al.* 2001). The *statistical sampling effect* models suggest that at higher diversities any given species in the regional species pool is more likely to have been 'sampled'. If a particular species has a great impact on an ecosystem process then systems with higher species diversity should on average have better functioning as there is a larger probability of this particular species being present in the community. The *niche complementarity* model, on the other hand, suggests that resources are used more efficiently at high diversities since the community consists of species with complementary (non-overlapping) niches as a result of differentiation due to interspecific, exploitative competition for limiting resources. The sampling model predicts that no higher-diversity community has a higher functioning, for example, primary productivity, than a monospecific community, whereas the niche complementarity model predicts that no one-species community is more productive than a two-species, no two-species more than three-species, etc. Recent empirical studies have supported the niche complementarity model (Kinzig *et al.* 2001).

Most empirical studies of diversity effects on ecosystem functioning have focused on primary producers in terrestrial grassland communities but there are some examples from freshwater systems as well. Studies on microbial microcosm systems have shown that increasing number of consumer species resulted in a strong negative effect on producer biomass (Fig. 6.7), that is, the ecosystem process consumption was highly dependent on biodiversity of the consumers (Naeem and Li 1998). Further, variability of biomass within functional groups decrease with diversity (McGrady-Steed *et al.* 1997) resulting in a higher predictability of ecosystem functioning. Wetlands provide important ecosystem services such as food production, recreational opportunities, and retention of nutrients. In many countries, wetlands affected by human activities are now restored and new ones are

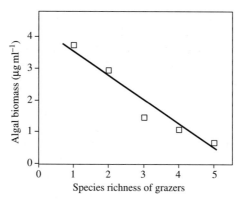

Fig. 6.7   The relationship between species richness of consumer species (flagellates and ciliates) and the biomass of producers (algae). From Naeem and Li (1998).

constructed, mainly to increase the retention of nutrients from the diffuse run-off from farmlands. Thus, it is important to maximize nutrient retention in these wetlands and it has been shown that retention of phosphorus increase with the diversity of submerged macrophyte species (Engelhardt and Ritchie 2001). A more diverse pond weed community was more efficient in physically filtering out phosphorus bound-up in particles. Nutrient cycling in lakes may also be affected by bioturbation, that is, mixing of the upper sediment layers by activity of organisms, including oligochaetes, amphipods, bivalve molluscs, and snails. They all differ in their use of the sediment as a resource and how deep they bury into the sediment suggesting that functional redundancy among species is very low and that species loss, that is, reduced biodiversity, affects the ecosystem processes involved (Mermillod-Blondin *et al.* 2002). Finally, if we take a short excursion to another freshwater habitat, it has been shown that decomposition of leaf detritus in streams increase with increasing species richness of detritivorous insects (Johnson and Malmqvist 2000) and although there of course are many differences between stream and lake systems it may not be too daring to predict the same pattern for lake detritivores.

Aquatic systems contrast terrestrial in that it is small organisms (i.e. algae) vulnerable to consumption that make up the major contribution to primary production. Further, consumer resource interactions are strong in freshwater systems and complex, top-down interactions are common (see Chapter 5, *Food web interactions*). Strong interactors (keystone species) high in the food chain may have cascading effects that control ecosystem process rates at lower trophic levels. Changes in the biodiversity of a relatively small number of key species may then have major effects on ecosystem functioning over relatively short timescales. Over the last decades a number of studies on interactions in pelagic and benthic food chains in freshwater systems have manipulated species composition or the presence/absence of trophic levels. Results clearly show that manipulating higher trophic levels have strong effects on important ecosystem properties such as biomass, primary productivity, water clarity, and nutrient retention/regeneration. Thus, the loss of biodiversity within one trophic level is likely to have large impacts on species at other trophic levels whether directly through changes in competitive and consumer resource inter-actions or indirectly via changes in ecosystem processes (Raffaelli et al. 2002). So human activities affecting biodiversity directly (hunting, fishing) or indirectly (e.g. pollution) may have profound influences on how the ecosystem functions. In some situations it may be necessary, or at least possible, to *restore* a system that has been severely disturbed by humans.

Restoration—to what?

Disturbances caused by large-scale processes such as climate change may not be reversible, but in many other situations it is possible to restore degraded lakes and ponds that have been subject to environmental disturbances. Because water is such an important resource for human populations, significant efforts have been made to restore disturbed freshwater systems to more pristine conditions. Acidified lakes and watersheds are limed, nutrient transport to highly eutrophic lakes is reduced, fish assemblages are manipulated to change the structure and dynamics of lake systems (bio-manipulation), and nuisance aquatic weeds are destroyed with biological control agents. However, when we try to restore disturbed systems we must be sure of what the ultimate goal of the effort is. In Europe new legislation, the Water Framework Directive, has as a goal that all lakes should meet the criteria for good ecological status and chemical quality by the end of 2015. However, then we need to know what 'good ecological status' is and this can only be defined if we know the background, pre-disturbance reference state for lakes of the same type as the one under consideration (Moss et al. 2002). But then we must know what characterized a pristine system and how far back in history we need to go to find such a state. For example, were lakes pristine before the industrialization started? Or before the change of the landscape when man went from being hunter/gatherers to farmers? One very useful tool for analysing the history of human impact on lake ecosystems and determining the characteristics of pristine lakes is to use paleolimnological methods.

Paleolimnology as a tool to understand history

In paleolimnology, the microfossil remains of organisms deposited in the sediment is analysed. Pollen and spores from terrestrial plants in the catch-ment area as well as remains of organisms that lived in the lake settle on the sediment surface and create a chronological record that could be used to recreate the historical environmental conditions of a lake. Pollen and spores of terrestrial plants and diatom frustules are the most abundant remnants in the sediment and they have been used to infer changes in climate (pollen) or nutrient enrichment and acidification (diatoms). For example, in recent decades there has been a dramatic increase in algal production in Lake Victoria, Africa, but it has been unclear what the relative roles are of increased nutrient loading following deforestation and intensified agriculture versus a reduction of planktivory by cichlids after the introduction of Nile perch (see Exotic species below). An analysis of the sediment record showed that diatom productivity was moderate up to 1930s but after that started to increase (Vershuren *et al.* 2002). In the late 1980s, the production of diatoms decreased dramatically as concentrations of silica, limiting diatom growth, were reduced. Burial of diatoms in the sediment depleted the lake's reservoir of dissolved silica. Instead, the algal community shifted towards a dominance of cyanobacteria. The sediment record shows that the increase in primary productivity in Lake Victoria started before the introduction of Nile perch and, thus, increasing nutrient levels is the major mechanism behind eutrophication.

Different remains of invertebrates in the sediment record may also be used to reconstruct the past history of a lake. Recent developments in the use of benthic and pelagic microcrustacean remains have proven particularly useful (Jeppesen *et al.* 2001). Some cladoceran zooplankton are hard-shelled and well preserved in the sediments. In other, more soft-shelled species only smaller fragments of the organism, such as claws and mandibles or resting eggs (ephippia), can be recognised. Changes in the abundance of invertebrate species or in the relative abundance of benthic versus pelagic microcrustaceans have been used to reconstruct the effects of water level fluctuations, contaminations, and invasive species. Plant-associated cladocerans have been used to infer changes in the dominance of submerged macrophytes in shallow lakes. Further, fish are size-selective predators and, thus, changes in the abundance of large *Daphnia* species versus small *Bosmina* indicate changes in the relative abundance of planktivorus fish.

In order to evaluate the microfossil record in the sediment we must first have a method to determine the age of sediment layers and, second, a method to relate the organism assemblage reconstructed from the micro-fossils to the environmental variables of the lake at the time when they were living. Several methods have been used to determine the age of sedimented material in lake deposits. The most important method of ageing sediment cores is to use isotope analysis. Organic material in older sediments (>400 years) can be aged by measuring its content of ^{14}C. More recent (<150 years) sediments can be dated by analysing lead (^{210}Pb) or cesium (^{137}Cs). In some lakes the sediments have very distinct layers due to seasonal differences in the rate of sedimentation and the quality of the sedimented material. This creates fine layers of alternating light and dark colour. When two layers represent one year the sediment is said to be *varved* and such sediments could be aged by just counting the varves. When the sediment layer has been aged we need a method to reconstruct the conditions of the lake at that time. A commonly used method is to construct so-called *transfer functions*. A transfer function is a multivariate statistical model that relates environmental conditions of a lake to organism remains in the sediment. For example, a transfer function relating diatom assemblages to total phosphorus concentration may be developed by taking water samples and sediment cores in a number of lakes with different phosphorus loadings. By analysing the species composition and abundance of diatoms in the top centimetres of the sediment core, which reflects the present diatom assemblage of the lake, and relating them to phosphorus levels one could develop mathematical expressions, transfer functions. These functions could then be applied to diatom assemblages from layers taken deeper down in the sediment core and be used to reconstruct the historical phosphorus concentration.

Even if paleolimnology provides a useful tool for characterizing what a pristine state is, it may still be difficult to determine the goal for a restoration project—should it be a return to the pristine state as defined by the paleolimnological analysis? Or should we accept, or even prefer, a system that is 'less pristine' but has a great value for recreational use, for high aesthetical reasons, for fisheries reasons, or for providing a good supply of drinking water? For example, considerable efforts have been made in Denmark to change shallow, eutrophic lakes from being dominated by phytoplankton to clear water lakes with lush stands of submerged macrophytes. However, recent paleolimnological studies have shown that the pristine condition was an oligotrophic lake with very few plants (N. J. Anderson and E. Jeppesen, personal communication). What was considered pristine in our limited time horizon was a system that had been eutrophicated by changing land use by early man. A return to the pristine stage is not feasible for the shallow Danish lakes of today, and possibly not even wanted. Lakes with well-developed submerged macrophyte vegetation and a diverse fish and invertebrate community may be more tractable. Below, we focus on some of the most serious threats to freshwater systems and look at measures that have been taken to reduce their effects.

Eutrophication

During the 1950s and 1960s, many lakes in urban areas and in areas with modern agriculture went through a process of drastic change. There are numerous stories told by older people of how a former pristine swimming lake within a few years turned into a waterbody with dense algal blooms, bad odour, and mucky bottoms. The reason for the change was not immediately identified, but the use of lakes as recipients for untreated sewage, as well as the use of agricultural fertilizers were among the suspected causes. Scientists suggested that it was the element phosphorus that specifically caused the *eutrophication process* in freshwaters. This suggestion was vigorously opposed, especially by the detergent industry, who were marketing products containing phosphorus as an active ingredient. Not until a whole lake experiment was carried out, was phosphorus finally accepted as being the major cause of the eutrophication process in lakes and ponds. In this experiment, a lake was divided into two parts and nitrogen (N) and carbon (C) were added to one of the basins while N + C + phosphorus (P) were added to the other. In the N + C basin, not much happened, whereas a magnificent algal bloom occurred in the basin to which phosphorus had also been added (Schindler 1974). The experiment showed that it was phosphorus, not carbon or nitrogen, that was the primary cause of the algal blooms so characteristic of eutrophicated lakes, contrary to what was claimed by the detergent industry and their experts.

The eutrophication process

The *eutrophication process* of lakes and ponds is characterized by *increased phosphorus concentrations* due to input, for example, of human sewage or fertilizer run-off from agricultural areas. This addition triggers a chain of events starting with a massive increase in the growth of primary producers since these are generally growth-limited by phosphorus in freshwater ecosystems. Periphytic (attached) algae and submersed macrophytes often show an increase in biomass during the beginning of the process (Fig. 4.9). However, phytoplankton, and especially cyanobacteria, soon take over as the dominant primary producers by reducing the amount of light penetrating down to periphytic algae and submersed macrophytes. The dominance of phytoplankton leads to a *reduction in water transparency*, and the high production of organisms results in an increasing amount of *dead organisms accumulating as sediment*. Bacteria mineralizing this organic material need large amounts of oxygen which they retrieve from the water, leading to a *reduced oxygen concentration*. This is especially pronounced in dimictic lakes during stratification when the thermocline prevents oxygen-rich surface water from reaching hypolimnetic water below the thermocline. *Fish kills* due to reduced oxygen concentrations are not uncommon in highly eutrophic lakes. However, the *amount of cyprinid fishes generally increases* considerably during a eutrophication process, which is illustrated by a survey of lakes differing in the degree of eutrophication (from 46 to 1000 µg phosphorus L^{-1}; Fig. 6.8). Since cyprinids are efficient predators on zooplankton, the abundance and size of grazing zooplankton decreases. Hence, the eutrophication process leads to considerable changes in the

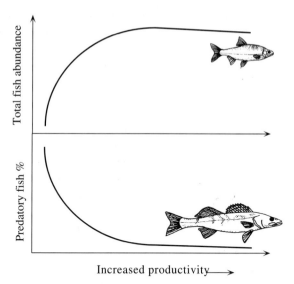

Increased productivity⟶

Fig. 6.8   The development of the total fish biomass in relation to the simultaneous decrease in the percentage of predatory fish at increasing productivity (eutrophication). Based on Jeppesen *et al.* (1996).

structure of the lake ecosystem, which in turn reduces the possibilities of using the lake for recreation, fishing, and as a source of drinking water.

Phosphorus reduction

When phosphorus had been identified as the cause of the eutrophication process in freshwaters, there was a consensus that the discharge of this element had to be reduced. The logical solution in this situation was to divert the waste water from the polluted lake or to clean it before it was released into the lake. The considerable investments in sewage treatment plants in many countries led to rapid reductions in the phosphorus concentration of many lakes, often resulting in considerable improvements in water quality. Surprisingly, however, far from all lakes recovered. Algal blooms still occurred and oxygen content in bottom waters, as well as water transparency was still low. Thus, phosphorus reduction through sewage treatment was not the universal answer to the eutrophication problem. The awareness of this was, of course, a disappointment that led to further research on other mechanisms that may have been responsible for the eutrophication process. One of the more important processes that prevented lakes from recovering was the huge amount of nutrients, in the form of dead organisms, that had accumulated as sediment during the eutrophication process. This sediment was very rich in nutrients which leaked to the water, a process called *internal loading* of nutrients. The internal loading should be distinguished from *external loading* which is the nutrient input from the lake's surroundings. Although sewage treatment and alterations in agricultural methods may deal with the external load, the internal load was predicted to remain high for decades, and even centuries, in many lakes. Hence, the discovery of the importance of internal loading called for restoration methods that could, if not revert the lake to its original state, at least reduce the problems.

Lake restoration

A number of technical solutions have been used to restore lakes and ponds, sometimes with good, but sometimes with less satisfactory results. We will briefly describe four methods: one that physically removes the sediment (*dredging*); one that uses chemical and bacterial processes (the *Riplox method*); one that is built on biological interactions (*biomanipulation*) and has its theoretical roots in food web theory. And one (*wetland construction*) that reduces nutrient input and at the same time may increase biodiversity.

Dredging

To remove the nutrient-rich sediment is, of course, the most logical solution to internal loading (Björk 1988). A dredger moves the sediment to artificial basins where the particles are allowed to sink. The water is either returned directly to the lake, or when the phosphorus concentration has been reduced with aluminium sulphate ($Al(SO_4)_2$) or iron chloride ($FeCl_3$). Phosphorus binds chemically to metals, such as aluminium (Al) and iron (Fe), and the resulting phosphorus–metal complex precipitates and can be removed mechanically. This method is also commonly used in sewage treatment plants. Due to technical and economic reasons, dredging is suitable mainly in small, shallow lakes and ponds.

The 'Riplox method'

The aim of this method is to reduce the internal loading of phosphorus by oxidizing the sediment surface, causing phosphate to precipitate in metal complexes. By pumping calcium nitrate ($Ca(NO_3)_2$) and adding iron chloride ($FeCl_3$) into the sediment, oxygen and iron (Fe) concentrations, respectively, are increased. The pH is stabilized with the addition of calcium hydroxide ($Ca(OH)_2$). At a suitable pH, denitrifying bacteria will transfer the nitrate in ($Ca(NO_3)_2$) to N_2 gas, which is released to the atmosphere (see Chapter 2, *The abiotic frame*). If all these steps work, an oxygen-rich 'lid' is created, reducing the release of phosphorus from the sediment into the water.

Biomanipulation

The term 'biomanipulation' was coined in the mid-1970s (Shapiro *et al.* 1975) and refers to the manipulation of biota in order to make a waterbody more desirable for humans. In general, this means reducing algal biomass in eutrophic lakes. Biomanipulations are generally performed by reducing the abundance of zooplanktivorous fish (usually cyprinids), either by addition of piscivorous fish, or by manually removing undesired fish (e.g. by trawling). The theory is that if the number of planktivorous fish is sufficiently lowered, the predation pressure on large zooplankton will decrease and the grazing rate on algae will increase. In this way, the likelihood of algal blooms will decrease and the water transparency increases (cf. Chapter 5).

Biomanipulation has been tested in several places in Europe and the United States with variable results. It appears that the best results are attained in shallow lakes, and when more than 80% of the initial biomass of planktivorous fish are removed Hansson *et al.* (1998b). However, the mechanism involved in the improvement is not necessarily only reduced predation on zooplankton, followed by higher grazing pressure on algae, and thus better light conditions. Instead, this mechanism mainly seems to be the trigger for other processes, including increased biomass of *submersed macrophytes* and *periphytic algae* at the sediment surface. Submersed macrophytes and periphytic algae absorb large amounts of nutrients which become unavailable for phytoplankton, but also *oxidize the sediment surface* which reduces the internal loading of phosphorus. Improvements may also be caused by *less disturbance of the sediment* surface by benthic feeding fish, as well as *reduced excretion* of nutrients by fish. It should also be noted that the *removal of fish* is also a removal of large amounts of phosphorus bound in their bodies, which would eventually have been released into the water once the fish died.

Wetlands as nutrient traps

A major part of the external input of nutrients to a lake often arrives through tributaries and drainage pipes from farmland. One way to reduce the nutrient loading to a lake is to construct wetlands in, or in connection to, these input sources. When incoming water from a stream or a drainage pipe is allowed to spread out on a larger area, such as a wetland, the water flow decreases and particles suspended in the water sink to the bottom. These particles are often nutrient rich and sedimentation is thus one mechanism through which wetlands reduce water nutrient concentrations. Submerged macrophytes also contribute to nutrient retention in wetlands. Wetlands are shallow and this allows high growth rates of macrophytes that incorporate large amounts of nutrients in their tissue. In addition, the vegetation increases the surface area available for colonization by denitrifying bacteria (see Chapter 2, 'Nitrification and denitrification') transferring nitrogen from the water to the atmosphere. Hence, wetlands remove nutrients from the water by *sedimentation, denitrification,* and direct *uptake by plants.* However, it should be noted that both denitrification and uptake in plants are temperature-dependent processes that are efficient mainly in the summer. There may, however, be additional values with wetlands, such as increased biodiversity of invertebrates, amphibians, and frogs (Zedler 2003).

Acidification

The words *acid rain* and *acidification* have gained much publicity ever since it was discovered that rain falling in highly industrialized areas had a low pH. Pure rain is slightly acidic (pH around 5.6), and, hence, acid rain is defined as rain with pH lower than 5.5. The reason for the low pH is that when we burn fossil fuel that contains sulphur, the sulphur is oxidized to sulphur dioxide (SO_2). When these oxides reach the atmosphere they react with photochemically produced ozone (O_3) and form sulfite (SO_3) which is dissolved in water droplets. The resultant sulphuric acid (H_2SO_4) lowers the pH of the rain considerably. The basic equations are as follows:

$$SO_2 + O_3 <-> SO_3 + O_2 <-> SO_3 + H_2O <-> H_2SO_4$$

Similarly, burning of nitrogen-rich fossil fuels creates several types of nitrogen oxides, such as nitrogen dioxide (NO_2), which form nitric acid (HNO_3) when dissolved in raindrops:

$$3NO_2 + H_2O <-> 2HNO_3 + NO$$

Effects of acid rain in freshwaters

The effect of acid rain differs widely among different waterbodies. In areas where the bedrock is rich in carbonates no effects of acid rain are recorded due to the buffering capacity of the carbonates; the lake is said to have high *alkalinity* (i.e. a high resistance to acidity). In areas with lower buffering capacity (lower alkalinity), acid rain may have a catastrophic impact on biota. Generally, there are dramatic drops in pH during periods with high rainfall or in spring after snow melt.

Changes in community composition due to reduced pH

When the pH in a lake or pond reaches values lower than 6, changes start to take place. Some of the effects may not be a result of the pH per se, but may instead be caused by secondary effects of the acidic environment, such as higher concentrations of aluminium and heavy metals (see below). Already at pH levels of 5–6, the *algal species diversity decreases* considerably, mainly because cyanobacteria and diatoms disappear. The algal flora becomes dominated by dinoflagellates (*Peridinium*) and chrysophytes (*Dinobryon*). The decrease in the number of algal species, as well as biomass, leads to *higher water transparency*; a characteristic feature of acidified lakes. In fact, the underwater photo on the front cover of this book is taken in a highly acidic and clear-water lake in Sweden. The number of periphytic (attached) algal species is also reduced, resulting in a dominance of a few acid-tolerant genera, such as the filamentous green alga *Mougeotia*, which

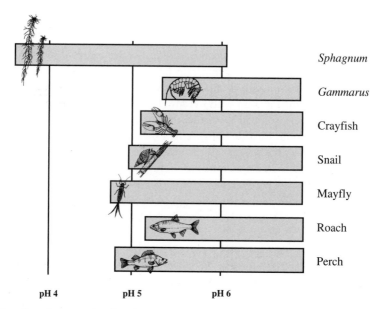

			Sphagnum
			Gammarus
			Crayfish
			Snail
			Mayfly
			Roach
			Perch
pH 4	**pH 5**	**pH 6**	

Fig. 6.9   The pH preferences showing that most organisms, with the exception of *Sphagnum*, prefer a pH above 6. At a pH between 6 and 5, animals, such as *Gammarus*, crayfish, and roach are lost, whereas mayflies and perch may reproduce in lakes with a pH as low as about 4.8.

may completely cover the sediment surface. Another acid-tolerant group is mosses, especially *Sphagnum*, which is a characteristic genus in acidified lakes and ponds (Fig. 6.9).

The fauna also becomes less diverse as acidification proceeds. Already in the pH interval 5–6, the *reproduction* of many animals is affected. When the pH drops to between 5 and 6, *Gammarus*, crayfish, and roach are lost from the system, whereas pike and perch, as well as mayflies, remain even at pH levels below 5 (Fig. 6.9). Generally, animals living in shallow parts of the lake are more affected by acidification due to the higher pH fluctuations and the acid flushes from the catchment area. Organisms living in, or on the sediment, however, are generally less affected since the sediment has a buffering effect. Among the zooplankton, daphnids are severely affected by acidification, while the abundance of other cladocerans, such as *Bosmina* (Fig. 3.13) may remain high. As the acidification process proceeds, the zooplankton community often becomes dominated by large calanoid copepods, such as *Eudiaptomus* (Fig. 6.10). The rotifers *Keratella* and *Polyarthra*, as well as some insect larvae, such as the phantom midge *Chaoborus* and corixids are often abundant in acidified lakes (Fig. 6.10). Further, one of the most significant features of acidified lakes is the disturbed reproduction of fish, resulting in fish populations being dominated by large, old individuals. As these individuals eventually die, the lake becomes fishless (Fig. 6.10).

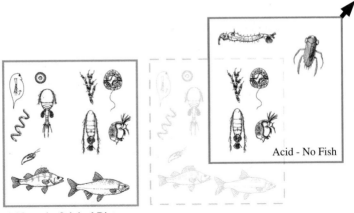

Neutral - Original Biota

Fig. 6.10   Some drastic changes in the organism community following acidification, that is, a change in the abiotic frame excluding some organisms and creating better conditions for other. Major changes are that fish, cyanobacteria, diatoms, *Daphnia*, and cyclopoid copepods eventually disappear, whereas, for example, calanoid copepods increase in importance. Invertebrate predators, such as *Chaoborus* and corixids, become the main predators. Based on Stenson *et al.* (1993).

Mechanisms behind changes in community structure

As we have seen, acidification causes dramatic changes in the community structure of lakes and ponds. The ultimate factor behind these changes may, however, not be solely the *low pH*, but also poisoning by *metals* (aluminium) or *heavy metals* (lead, cadmium), which become more soluble in acid environments. Fish are especially sensitive to aluminium since this metal binds to the gills and affects their respiration efficiency. Another mechanism is the reduction in phosphorus concentration probably due to chemical binding and precipitation with aluminium. This *oligotrophication process* may lead to nutrient depletion among many organisms. Moreover, the disappearance of fish releases many invertebrates from a heavy predation pressure. It is likely that the high abundance of large predatory invertebrates, such as *Chaoborus* and corixids, in acidified lakes is a result of the *release from predation and/or competition* from other organisms.

Actions against acidification

In the 1970s and the 1980s, there was a widespread recognition of the damaging biological effects of acid rain on freshwater systems and a strong public opinion developed. This resulted in political decisions in both Europe and North America for actions to reduce the problem. Management actions for the recovery of acidified freshwater systems operate at different spatial and temporal scales. On the large scale, across regions and national boundaries, changes in policies and legislation have resulted in reduced emissions of acidifying substances, that is, a management action directed towards the *cause* of the problem. On the local scale the negative effects of acidification have been circumvented by liming, that is, addition of calcium carbonates ($CaCO_3$). Liming is a management option that gives instantaneous improvements but it only treats the *symptoms* of the problem.

Reduced emissions

The sustainable measure against acidification is of course to reduce emissions of acidifying substances. However, since reductions in emissions include economic decisions this generally requires tedious negotiations, often on an international level. Fortunately, such negotiations were initiated already during the 1970s in Europe and North America. Legislation that have promoted new emission control techniques and a reduction in the use of sulphur-rich fuels have resulted in considerable reductions in sulphur emissions in Scandinavia (Fig. 6.11) and North America, which have resulted in increased pH in rainwater (Stoddard *et al.* 1999). In spite of the significant decreases in deposition of sulphuric acids only slight improvements of pH have been detected in most acidified systems (e.g. Gunn and Keller 1990, Stoddard *et al.* 1999, Jeffries *et al.* 2003). After a *chemical recovery* of a lake system the chemical conditions are so favourable that they allow survival and reproduction of acid-sensitive plants and animals, that is, there will be a *biological recovery* of the system. The rate of recovery is dependent on the tolerance limits of the organisms to acid conditions but is also due to their colonization rates; zooplankton likely recover within 3–10 years, whereas fish populations need another 5–10 years to recover (Driscoll *et al.* 2001). The slow chemical recovery of acidified lakes even though sulphur emissions have been dramatically reduced is probably due to a continuously high deposition of NO_x. In Sweden there has been a slight reduction since 1990 (Fig. 6.11) whereas in the United States emis-

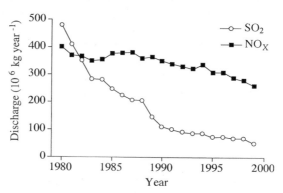

Fig. 6.11   The discharge of SO_2 and NO_x over Sweden from 1980 to 1999, showing a clearly decreasing trend with respect to SO_2, but only a slight decrease in NO_x. (*Source:* Swedish Environmental Protection Agency.)

sions of NO_x have remained fairly constant (Driscoll *et al.* 2001). The lower rate of emission cuts of NO_x is mainly due to increasing emissions from vehicles but also from industry and agriculture. Earlier studies on nitrogen deposition focused primarily on the acidifying effects but recently there has been increasing research on the role of atmospheric nitrogen deposition as a nutrient source in freshwater systems. Phosphorus is traditionally regarded as the limiting factor for lake productivity but recent studies have shown that nitrogen is the primary or co-limiting nutrient for phytoplankton growth in many unproductive lakes (Jansson *et al.* 1996; Maberly *et al.* 2002). Thus, besides acidification a continued high emission rate of nitrogen may affect lakes also by increasing eutrophication.

Liming

A fast measure of making an acidified lake return to its former pH is to add lime (calcium carbonate; $CaCO_3$) to a specific lake or watershed. This has been extensively used as a restoration tool in Scandinavia. In Sweden, it is estimated that more than 17 000 lakes are affected by acidification. More than 7000 of these lakes, which account for 90% of the area of acidified lakes in Sweden, have been treated with lime during the last 10–15 years at a high annual cost (about 20 million Euro year^{-1}) and it is expected that liming will have to continue for another 10–20 years. Liming leads to a rapid increase in pH, but the improvements following lime addition eventually vanish, meaning that repeated treatments are necessary. A recovery of the chemical conditions following liming also results in a biological recovery, that is, an increase in species diversity and biomass of most organisms. Phytoplankton, zooplankton, macrophytes, and invertebrates increase rather rapidly in limed lakes due to natural colonization (Weatherly 1988) whereas recovery of fish is somewhat slower. However, a study of acidified lakes showed that fish species richness increased after liming and was of the same magnitude in limed lakes as in naturally circumneutral lakes after 10–20 years (Appelberg 1998). A re-establishment of the fish fauna was accelerated by deliberate introductions of fish, as colonization rate was a limiting factor. Although liming has been relatively successful as a local and regional restoration method for specific lakes and ponds, it certainly does not solve the acidification problem. Acid rain continues to fall on the ground, on the sea, and on lakes and ponds.

Contamination

Since water is an efficient solvent for many substances, lakes and ponds have since ages been used for disposal of domestic, agricultural, and later, industrial wastes. Besides organic matter and nutrients causing oxygen deficits and eutrophication, these wastes have also contained substances that are toxic in one way or another to the organisms living in the ponds and lakes. The most important contaminants in freshwater systems are heavy metals and organochlorine substances. They have some common features that make them serious threats to freshwater biota. Both are persistent pollutants in that they are not broken down (metals) or take a very long time to degrade (organochlorines) so they effectively become permanent additions to the lake or pond ecosystem. Some of these substances, but not all, accumulate in the food chain with top carnivores having larger concentrations of the contaminant than organisms at lower trophic levels. They also bind to particles that fall out of the water column and become incorporated in the sediment. The sediment acts as a sink for these contaminants which can accumulate in large quantities. Sediment-dwelling organisms, such as mussels, oligochaetes, and chironomids, are thus more exposed to contamination and commonly have high concentrations of heavy metals and/or organochlorines. Biological activity of sediment dwelling organisms may in turn make heavy metals and organochlorines bioavailable, either as their activity expose them to the overlying water again or in that sediment organisms are prey for invertebrate predators or fish from the pelagic habitat. Heavy metals and organochlorines have a number of effects on organisms, including direct toxic effects and longer-term effects such as carcinogenic effects, neurological disorders, reduced growth, disabled immune system, and reproductive disorders. Lately, it has also been recognized that male and female hormone systems may be affected by contaminants, resulting in reproductive failures (see endocrine disruption below).

Heavy metals

Heavy metals are natural substances that enter the lake and pond systems through natural weathering of rocks, from volcanic activities or, more seriously, through anthropogenic activities such as mining, smelting, different industrial processes, burning of fossil fuels, and refuse incineration. Although problems with heavy metals in lake systems often are the result of point source pollution, high levels of heavy metals, such as mercury, in otherwise pristine lakes suggests that atmospheric transport may be important in some cases. Many metals are essential components of physiological processes in organisms, but at higher concentrations they all become toxic. The metals being the most serious threats as environmental contaminants include mercury, cadmium, and lead. Surveys of heavy metal concentration in lakes have shown that metal concentrations in organisms are correlated to environmental variables of the lake and its watershed, such as pH, dissolved organic carbon (DOC), lake trophy, land use, temperature, and they differ in their propensity to biomagnify along the food chain (Chen *et al.* 2000).

Organochlorines

Chlorinated organic substances used as pesticides and in industry have for long been recognized as a major environmental problem in lakes and ponds due to their effect on the biota and their resistance to chemical and biological (microbial) breakdown. Organochlorines are hydrophobic, fat-soluble, and biologically stable compounds that accumulate in the body fat of freshwater organisms. Some pesticides have been applied directly to freshwater systems to control populations of freshwater organisms, for example, DDT (dichloro-diphenyl-trichloroethane) used for control of malaria mosquitoes, but the major source of pesticides to lakes and ponds is diffuse run-off from terrestrial systems where organochlorine pesticides have been and are used in agriculture. Polychlorinated biphenyls, or PCBs, is another important class of organochlorine contaminants. PCBs are stable liquids of low volatility and high stability that have been used in industry as for example, hydraulic fluids, lubricaters, insulation fluids in transformers and as plasticizers in paints. Their high resistance to biological degradation has eventually resulted in an accumulation in freshwater systems where in some cases they have had strong adverse effects on the biota. Today, atmospheric fall-out is the dominating input of organochlorines in lakes and ponds (Muir *et al.* 1990) resulting in that organism in pristine lakes in remote areas still have significant concentrations of organochlorines. In Alaskan lakes, it has also been shown that salmon migrating from the sea to spawn transport large amounts of PCB and DDT, accumulated during their ocean life-stage. This is then deposited in the spawning lakes and accumulated in the food web (Ewald *et al.* 1998). In these lakes such biotransport is a more important source of persistent organochlorines than atmospheric fall-out. A reduction of organochlorine usage resulting from restrictions enforced in most industrialized countries during the 1970s has led to a reduced contamination of the aquatic environment. Since then there has been declining concentrations of, for example, PCBs and DDTs in fish (Fig. 6.12), mussels, and other freshwater organisms, with a more rapid decline in some substances (e.g. DDTs) than in others (e.g. PCBs) (Loganathan and Kannan 1994). Most studies have shown that local restrictions on the use of organochlorines have had positive effects on local freshwater environments, but atmospheric fall-out will continue to affect lake and pond ecosystems.

Endocrine disruptors

In recent years, there has been an increased interest in the importance of endocrine disruptors in aquatic systems. An endocrine disruptor is a hormone-like substance that changes the endocrine function and causes harmful effects on the organisms, its offspring, and/or has negative

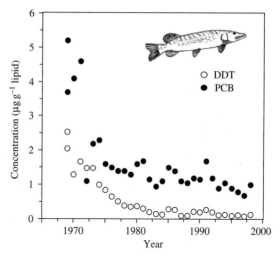

Fig. 6.12   The concentration (μg g^{-1} lipid) of DDT and PCB in muscle tissues of northern pike (*Esox lucius*) from Lake Storvindeln, Sweden. (*Source*: Swedish Environmental Protection Agency.)

effects on exposed populations of the organism. Of particular interest are estrogenic substances that mimic steroid hormones. Aquatic organisms seem to be especially affected by endocrine disruption (see Mathiessen and Sumpter 1998 for a review). It was discoveries of morphologically abnormal male sexual organs in alligators, feminization of male fish and turtles, and masculinization of female fish in process water from pulpmills that first drew attention to the problem. Suggestions that endocrine disruptive compounds may be responsible for decreased sperm counts and increased testicular and breast cancers in humans have created a huge interest in the subject. Studies in England showed that male roach in waters downstream outlets from sewage treatment plants had signs of feminization expressed as development of female sexual organs in parallel to the male organs, that is, resulting in hermaphroditic fish. The degree of feminization was related to the load of polluted water and it was suggested to be due to remains of natural estrogenic hormones and synthetic oestrogen from contraceptive pills in the effluent water. Laboratory studies have later shown clear effects of these substances on different aspects of the reproduction cycle in fish. Other substances that are similar in structure to oestrogen have also been shown to have endocrine disruptive effects, including substances such as chlorinated insectides and their degradation products, phthalates, dioxines, PCB, and others. Further, these substances may act synergistically. However, although refined analytical methods have shown effects from endocrine disrupters in many species in a range of different aquatic habitats, it is still not clear to what degree this is a threat to their populations or to the aquatic ecosystems (Mathiessen 2000).

Global climate change

Temperature increase

Global warming is another environmental problem that has received substantial attention lately. Anthropogenic emissions of gases, such as carbon dioxide (CO_2), form a layer in the atmosphere that reduces the amount of heat leaving earth, leading to increased mean temperatures. Although there has been an intense debate on whether there is a global warming caused by our burning of fossil fuel or not, the general consensus today is that there exists a causal relationship between temperature change and green house gases. Instead of discussing if there is an effect, the scientific discussion now focus on the magnitude of a temperature increase and what the effects on ecosystems will be. In lakes, increasing levels of greenhouse gases and the likely increase in water temperature will lead to changes in seasonal temperature patterns affecting, for example, stratification and ice-out. Long-term datasets have already shown reduced duration of ice-cover (Fig. 6.13). Regional hydrology, such as reductions in streamflow and lake water levels, may also be affected. Moreover, changes in lake volume and thermal structure, as well as alterations in catchment inputs of detritus and nutrients and an increase in the frequency of extreme events such as droughts and floods may be expected (Sala *et al.* 2000). Hence, synergistic, and often non-intuitive, effects between a possible increase in temperature and changes in the hydrological regime of the catchment, may have profound effects on the amount of sediment and nutrients reaching lakes and ponds.

Changes in the abiotic frame of lakes and ponds induced by global warming, such as changes in nutrient input and heat loading may have strong effects

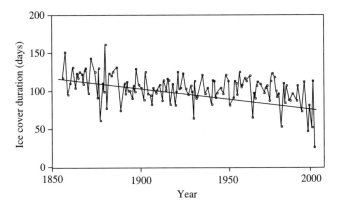

Fig. 6.13   The duration of ice cover on Lake Mendota, Wisconsin, USA since the mid-1800. From Kling *et al.* (2003).

on the biota, altering the species composition of phytoplankton, zooplankton, benthic invertebrates, and fish (Magnusson *et al.* 1997). Experiments on microbial food webs have shown that warming can cause a reduction in species richness with a higher extinction rate of herbivore and predator species (Petchey *et al.* 1999). The loss of species at higher trophic levels resulted in increased producer biomass and primary productivity. In contrast, primary productivity and fish production has decreased in Lake Tanganyika, Africa, as a result of increased water temperatures during the last century (O'Reilly *et al.* 2003). Increased water temperatures and reduced wind velocities have resulted in reduced mixing of the water column, decreasing upwelling of nutrients from deep water to surface water and this has resulted in a reduction of productivity of the oligotrophic lake. On a larger scale, it has been shown that climate may have strong effects on interactions in lakes and ponds. The climatic phenomenon called the *North Atlantic Oscillation* (NAO) affects food web interactions in central European lakes (Straile 2002). The NAO affects winter climate variability in Europe, and is strongly correlated to the duration of the *spring clear water phase*, a period in spring characterized by unusually low biomass of phytoplankton. The correlation is due to that spring water temperatures control the development of zooplankton, and in this case *Daphnia* is especially important, that increase in biomass. The high grazing pressure on phytoplankton results in a period of clear water. Temperature is also of fundamental importance for the life history of aquatic organisms affecting, for example, metabolic and development rates. Hence, an increase in temperature may affect the hatching date with far-reaching effects on size at hatching, food availability, and over-winter survival (Chen and Folt 1996). Kolar and Lodge (2000) have suggested that such changes in ecosystem structure due to global warming will make the system more vulnerable to invasions by exotic species.

Increasing temperature has also been predicted to result in that the distribution of vector-borne diseases, such as malaria (caused by *Plasmodium falciparum* carried by mosquitoes), should expand into regions that are at present too cool for their persistence (Martin and Lefebvre 1995). Recent model predictions forecast, however, that the distribution of malaria will change little, despite an increase in temperature (Rogers and Randolph 2000). The argument for the latter scenario is that the interactions between host and vector are so complex.

Solar ultraviolet radiation

The sun has sent radiation towards the earth long before any life appeared here, but the formation of an ozone layer (O_3) in the upper part of the atmosphere, when oxygen became a significant part of the atmosphere, has reduced the amount of ultraviolet (UV) radiation reaching earth. Recently, it was discovered that our use of certain chemicals (chlorofluorocarbons, CFCs), caused the ozone layer to thin. In the stratosphere, the CFCs react with UV radiation and atomic chlorine is released. The highly reactive chlorine then splits ozone into one O_2 and one O. The chlorine is not affected by the reaction, but continues to attack new ozone molecules. The consequence is that 'holes' in the ozone layer are created causing higher amounts of UV radiation to reach the earth's surface, especially at high latitudes. High amounts of UV radiation are very harmful to organisms and affect many of the fundamental cell processes, including DNA replication and cell metabolism.

The UV radiation attenuates rapidly through the water column (e.g. Williamson *et al.* 1996), and only a small proportion of the most harmful part that reaches the surface of the earth, namely UV-B radiation (280–320 nm wavelength), penetrates deeper than 8 m in clear water. In humic lakes, where the amount of DOC is high, it may only reach a depth of a few centimetres (Williamson 1995, Schindler *et al.* 1996, Fig. 6.14). However, exposure to UV-B radiation may result in a breakdown of recalcitrant DOC to smaller molecules that can be used by bacteria and thus lead to a higher bacterial production (Williamson 1995; Lindell *et al.* 1995). However, the reaction between UV and DOC may also result in the formation of harmful substances, such as hydroxide radicals and hydrogen peroxide (H_2O_2), which may inhibit bacterial production (Xenopoulos and Bird 1997).

The sensitivity to UV radiation differs among organisms, for example, biomass of benthic algae was reduced when exposed to high levels of UV radiation, but in the presence of benthic grazers (chironomid larvae), the patterns were reversed with benthic algae having greater biomass at the higher UV radiation (Bothwell *et al.* 1994). The likely explanation for these results is that although the benthic algae were negatively affected by UV radiation, their grazers (the chironomids) were even more sensitive. In

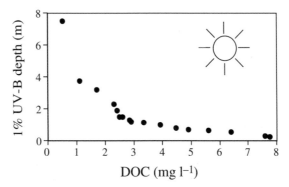

Fig. 6.14   The relationship between DOC and the depth to which 1% of UV-B radiation reaches in some Canadian lakes. Data from Schindler *et al.* (1996).

this way, UV radiation and other environmental hazards may indirectly affect the equilibria between different groups of organisms, such as consumers and producers. Another notable example of some organisms being better adapted to UV radiation is that certain zooplankton groups have pigments protecting them from harmful radiation; in *Daphnia*, the pigment melanin functions as a sun screen (Hill 1992). Copepods, on the other hand, have the red pigment astaxanthin which functions as an antioxidant by neutralizing carcinogenic substances induced by UV radiation. Interestingly, copepods seem to adjust their level of pigmentation in relation to the prevailing risk; being highly pigmented but vulnerable to predators, or being transparent but unprotected against UV radiation (Hansson 2004). Hence, some aquatic organisms seem to be better adapted than others to a possibly increasing UV influx, and such organisms may be predicted to play a more important role in future ecosystems. Increase in UV radiation may therefore lead to considerable alterations in the composition of aquatic communities, especially in shallow, clear-water lakes and ponds. However, effects of increasing UV radiation will most probably be less catastrophic in aquatic than in terrestrial systems, since the water surrounding the organisms absorbs much of the UV radiation.

Combined effects of environmental hazards

As we have seen, the influence of each of the global environmental hazards on ecosystem processes is considerable. Even more serious, and less fully appreciated, are the effects of combined environmental problems, which may lead to non-intuitive synergistic responses. For example, the increase in temperature has led to a drier climate, a reduced stream flow, and thereby, a reduction in the export of DOC from terrestrial habitats to lakes and ponds (Schindler *et al.* 1996). The lower DOC input has resulted in clearer lakes where the UV radiation penetrates deeper. An example of how such synergies between climate-induced drought and increased UV radiation affect freshwater organisms can be taken from amphibian reproduction. Reduced water level at amphibian oviposition sites caused an increased embryo mortality due to greater exposure to UV radiation, followed by infections by the fungus *Saprolegnia ferax* (Kiesecker *et al.* 2001). Hence, higher temperature led to lower water levels, exposing the amphibian embryos to an increasing intensity of UV radiation, which, in turn, reduced resistance to fungal infection. Thus, environmental hazards may act together to strengthen an effect, but may also lead to completely unexpected consequences. There are, unfortunately, reasons to believe that the future will bring additional unexpected events caused by such synergistic effects among environmental problems; effects that are almost impossible to predict.

Timescale of environmental change

On an evolutionary timescale (thousands of years), the species composition and dynamics of lake and pond systems have not been constant. Instead, there have always been changes due to natural environmental oscillations, such as climate changes, volcanic eruptions, and other catastrophes. One example of drastic change is the bacterial mutation that allowed cyanobacteria to start photosynthesizing, which 'poisoned' the earth's surroundings with oxygen and made completely different organisms dominant. Another example is the unknown environmental factor that caused dinosaurs to be replaced by mammals. The difference between these examples and the current environmental changes is the *short timescale* on which today's changes occur; 50 years compared to many thousands of years. Moreover, the environmental changes we have discussed in this chapter have all been induced by one species: *Homo sapiens*! Another important aspect is that *we are aware* that we are destroying our environment and thereby the possibility of our continued existence on earth!

Exotic species

In later years, increased travel and international commerce has resulted in a start of a global homogenization of our biota through the dispersal of organisms outside their native ranges. In lakes and ponds, this invasion of exotic species has in many cases resulted in loss of biodiversity and changes in community structure and ecosystem functioning. However, far from all organisms colonizing habitats outside their native range have dramatic effects on their new environment. Most colonizers to new systems do not survive for long or are kept at such low population densities that they do not affect the system noticeably. Further, invasive species commonly go through a lag phase from establishment to becoming an invader and only a small fraction of the colonizing species actually become invasive, having dramatic impacts on the invaded system. Aquatic habitats are particularly sensitive to invasions by exotic species because dispersal between freshwater systems is facilitated by human activities, but also because dispersal within a water system is rapid and often relies on passive transportation.

A successful invasion of an exotic species has dramatic effects on different organizational levels of the invaded system (Parker *et al*. 1999). At the individual level, invaders may cause behavioural changes such as habitat shift in native species due to competition or predator avoidance, and thereby cause reduced growth or reproduction. Genetic effects may occur as a result of changes in natural selection or gene flow or due to hybridization. At the population level, invaders may cause changes in size and age structure, distribution, density, population growth, and may even drive populations to extinction causing reductions in biodiversity at the community level. In fact, more than 40% of the species listed as endangered or threatened in the United States are considered to be at risk due to competition or predation by exotic species (Pimentel *et al.* 2000). At the ecosystem level, invasion of exotic species may result in changes in nutrient dynamics, rates of resource acquisition, disturbance regimes, and it may also change the physical habitat affecting biotic processes. Naturally, all these changes have dramatic impacts on the biota of lakes and ponds, but invaders also cause huge economic losses including losses of goods and services as well as costs for control. In the United States only, direct costs of exotic aquatic weeds, fishes, and zebra mussels was estimated to $110, $1000, and $100 millions per year, respectively, and that was excluding values associated with species extinctions, ecosystem services, and aesthetics (Pimentel *et al.* 2000). Thus, identification of future invasive species and potential sites of invasion would be of great value, both from an environmental and economical point of view. Successful invading exotic species have been suggested to have certain traits in common, including characteristics such as fast growth, short reproductive cycles, generalists with wide environmental tolerance, broad diets, and having mechanisms of efficient dispersal (e.g. Ricciardi and Rasmussen 1999). However, this may be an oversimplification as quantitative analyses provide little support to such generalizations of invasive traits. Instead, researchers now adopt a more integrated approach where characteristics of both the invader and the invaded community are included. As mentioned above, ponds and lakes could be seen as freshwater islands in a 'terrestrial ocean' and it has been suggested that islands are more at risk of being victims to ecological damage due to invasion of exotic species (e.g. Simberloff 2001). Further, the biodiversity of a community may also affect its resistance to invasion by exotic species. Species rich communities are more likely to have species that can exclude invaders through competition, predation, or parasitism. According to this 'biotic-resistance' model species-poor communities should be very vulnerable to invasion but as new species are incorporated in the community the rate of establishment should decrease (Fig. 6.15). An alternative hypothesis, the 'invasional meltdown' model (Simberloff and Von Holle 1999) instead suggests that the rate of invasion increases with cumulative number of successful invaders (Fig. 6.15). This could result if there are positive interactions among introduced species, for example, that early invaders change the habitat in some way so that later invaders bene-fit. Early invaders may also be preferred prey of late invading predatory species. Experiments in small-scale systems have provided evidence for the biotic-resistance model. For example, establishment of introduced zooplankton species to experimental pond systems was lowest in ponds with highest initial diversity of zooplankton, suggesting that diversity increases resistance to invasion (Shurin 2000). In contrast, an analysis of the invasion history of the Great Lakes of North America suggested that the

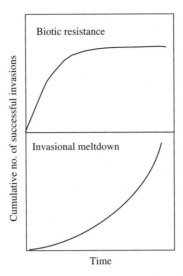

Fig. 6.15   The change in the cumulative number of successful invasions over time as predicted by two models, the biotic resistance model and the invasional meltdown model. From Ricciardi (2001).

cumulative number of invaders follows the 'invasional-meltdown' model and, further, that the majority of interactions among established invaders were positive, suggesting facilitation between invading species (Ricciardi 2001).

Exotic species may be introduced to the system intentionally, for example, as a part of a stocking programme to increase fisheries yields, in aquaculture systems and, historically, when settlers on new continents have brought their homeland species to their new countries. In recent years there have been an increase in unintentional introductions. The building of large canals that connect large water systems has facilitated the dispersal between systems and, further, ships navigating in these systems dump ballast water, which may contain millions of freshwater organisms from other continents. The Great Lakes of North America are good examples of systems that have suffered greatly from the invasions of exotic species. In some cases the new species have caused substantial economic damage (see Mills *et al.* 1994 for a review on the impact of exotics on the Great Lakes). A dramatic European example is the spread of the North American crayfish plague in the early 1900s, which wiped out the indigenous crayfish species and caused a severe decline in crayfish fisheries.

In order to understand the effects of exotic species and how new invasions are stopped or at least reduced, more research is needed on how invaders affect native species and why some species not only establish in a new habitat but disperse explosively and build-up huge populations that have considerable impact on the native communities. By studying the dramatic effects of exotic species we may also learn a great deal about the factors that are important for the structure and dynamics of natural communities (Lodge 1993). Below, we deal with three examples of invading organisms that have had a tremendous impact on their new system: freshwater weeds, zebra mussels, and Nile perch.

Freshwater weeds

At times, an aquatic plant may become so abundant that it becomes a nuisance—a weed. Aquatic weeds are typically species that have spread outside their original geographical distribution area. Ironically, plants that are considered as weeds in some parts of their distribution may be desired or even endangered in their native area. For example, large efforts are being made in Europe to restore submerged vegetation in shallow eutrophic lakes devoid of plants (see the section on biomanipulation above), whereas submerged macrophytes are generally treated as nuisance plants in large parts of the world. Weeds typically have rapid vegetative growth and a high dispersal rate. In many places they seriously interfere with human activities, for example, by hindering navigation, reducing water movement in irrigation or drainage canals, disturbing the operation of hydroelectric power plants, and decreasing the fish population.

In Europe, the submerged plant *Elodea canadensis* spread over the continent during the latter part of the nineteenth century and in later years eutrophication has led to increased stands of the common reed *Phragmites australis* in lowland lakes. In North America the Eurasian water milfoil (*Myriophyllum spicatum*, Fig. 3.4) has caused severe problems. However, the most spectacular and serious examples of the spread of aquatic weeds come from subtropical and tropical areas. Typically, floating weeds are troublesome because they interfere dramatically with human activities. Two of the most widespread and serious weeds are the water hyacinth (*Eichhornia crassipes*) and the floating fern (*Salvinia molesta*). Their high growth rates (doubling time of less than 14 days at optimal conditions) result in the build-up of large plant biomasses that are choking huge areas of freshwater habitats in the tropics.

Control methods for freshwater weeds

Of course, significant resources have been invested in various weed control programmes and a number of methods have been developed to get rid of freshwater plants. The simplest, but also the most labour-intensive, is *manually removing* the weeds. Different mechanical devices have been developed to increase the removal efficiency, from small-scale weed-cutting boats to large harvesters. *Herbicides* have been a cheap, effective, and rapid method in some cases, but then there is always a problem with unwanted toxic effects on non-target organisms, such as non-weed plants, aquatic invertebrates, fish, and in some cases, humans. *Biological control* agents have been used as an alternative in many parts of the world. A potential control agent is typically a specialized herbivore on the plant, for example, a phytophagous insect feeding on the plant in its native range, but not occurring in the region where the plant has been introduced. Other effective control organisms are generalist herbivores such as grass carp, snails, or even manatees. The grass carp (*Ctenopharyngodon idella*), native of rivers in Siberia and northeast China, has been successfully used as a control agent in large parts of the world. It is a voracious grazer on aquatic plants and can tolerate a wide range of environmental conditions, although it will not breed naturally outside its native habitat. Thus, populations of grass carps can easily be kept under control. However, releasing these generalist herbivores under other than highly controlled conditions may be risky; they can spread and be a threat to other plants, such as rice crops.

Perhaps the most well-documented case of biological control is the *Salvinia* example (see Room 1990 for a review). *S. molesta* has spread through much of the subtropics and tropics, forming thick mats, up to a metre in depth, that cover lakes and slow-flowing rivers. The plant was first described as another species and it was not until the beginning of the 1970s that it was correctly described as *S. molesta*. The problem was that nobody knew its native range and thus potential biological control agents could not be found. When the geographic origin of *S. molesta* was discovered in southeastern Brazil, three different herbivores that fed on the plant were discovered—a weevil, a moth, and a grasshopper. The weevil, which feeds on buds and roots, was introduced into an Australian lake as a potential biological control agent and within a year it had increased from a population of a few thousand to more than 100 million individuals and removed some 30 000 tons of *Salvinia*! The weevil is now used as a biological control of this plant throughout the tropics and typically reduces its populations by 99% within one year.

The zebra mussel

In 1988, an exotic mussel was discovered in Lake St Clair, North America—the zebra mussel, *Dreissena polymorpha* (Fig. 3.11). Within a few years it had exploded in density and spread throughout the Great Lakes and large parts of the Mississippi River watershed (Ludyanskiy *et al.* 1993). This little mussel has turned out to be one of the most destructive and expensive invaders in North American freshwater systems, causing million dollar costs by latching on to any hard surfaces it can find, from other mussels to water intake pipes. Where did this mussel come from and why has it been so successful?

The zebra mussel is native to the Black and Caspian seas and during the nineteenth century it spread throughout European inland waters. It seems likely that the mussel 'hitch-hiked' across the Atlantic in a ship, which dumped its ballast water in Lake St Clair sometime in 1986. The zebra mussel has several features that made it an excellent invader: it attaches to hard substrates with byssus threads and this was an underexploited or vacant niche for large filter-feeding organisms. Moreover, it has a high production of planktonic larvae (up to 40 000 per female), and thus can disperse rapidly within a waterbody. Finally, there were no, or few, predators, parasites, or diseases in the new environment.

Besides having an effect on water intake pipes and other human constructions, it soon became clear that the extremely high densities (in some places more than 700 000 individuals m^{-2}!) could have drastic ecological effects, including competitive exclusion of other mussels and changes in energy and nutrient flows of the ecosystem. Native unionid clams were covered by zebra mussels, affecting feeding, respiration, and locomotion of these organisms. In systems with high densities of zebra mussels the unionids have declined dramatically and many species have disappeared locally. There was also great concern that the heavy infestations of zebra mussels on the offshore bedrock reefs would affect population densities of fish that used these reefs as spawning grounds, such as the economically important walleye (*Stizostedion vitreum*). However, it was later shown that mussel densities of up to 334 000 m^{-2} had no adverse effect on walleye egg deposition, egg viability, or oxygen availability (Fitzsimons *et al.* 1995). In recent years, a sibling species, the quagga mussel (*Dreissena bugensis*) has started to appear. The quagga mussel can colonize and survive also on soft substrates such as sand and mud and it is also found at greater depths than the zebra mussel (Bially and MacIsaac 2000).

The great densities of this efficient filter-feeder suggest that it may have a considerable effect on the biomass of phytoplankton. Theoretical estimates based on clearance rates and bioenergetic modelling have shown that reef-associated zebra mussels in western Lake Erie may filter 14 times the volume of the basin, and thereby remove phytoplankton from the water column to an amount equivalent to 25% of the primary production—each day! In several lakes the water clarity has actually increased after the invasion of the zebra mussels and, as a result of better light conditions submerged macrophytes have increased in abundance. In The Netherlands zebra mussels are in fact used in programmes to decrease the negative effects of eutrophication. By efficiently filtering phytoplankton, zebra mussels are shunting carbon and nutrients from the pelagic to the benthic zone through the build-up of mussel biomass, but also by the production of faeces and pseudofaeces which enrich the sediment and change the living conditions for many benthic invertebrates. Besides sediment enrichment, zebra mussels also benefit benthic macroinvertebrates by increasing structural complexity and thereby increasing available substrate and providing a refuge from predation (Stewart *et al.* 1998).

Nile perch

The Great Lakes of East Africa (Lakes Victoria, Tanganyika, Malawi) are famous for having an exceedingly diverse fish fauna (Kaufman 1992). In Lake Victoria, the largest freshwater lake in the world, a species flock of cichlids dominated the fish fauna both with respect to number of species and biomass. This monophyletic species flock consisted of over 300 species which were thought to have radiated over the last 750 000 years. The majority (>90%) of these species were endemic; they were only found within the lake and some species even showed very restricted distribution ranges within the lake. These cichlids were very similar in appearance and morphology, but they had evolved a remarkable set of trophic specializations that enabled them to make use of practically all different types of food sources found in the lake, from detritus to fish.

Increased fishing pressure due to expanding human populations and more efficient fishing methods resulted in overfishing of the large *Tilapia* species and a drastic decline of fish yields in the late 1950s. There was still a large stock of cichlids, which was of considerable importance for local inhabitants, although not of interest for industrial fishing. To improve the possibilities for commercial fishing, Nile perch (*Lates* spp.) from Lake Albert and Lake Turkana were introduced into the northern parts of Lake Victoria in the early 1960s. The aim was to convert the large biomass of small, unpalatable cichlids to a suitable table fish. The Nile perch is a voracious predator with a considerable plasticity in foraging habits, feeding on a wide range of prey types and sizes. During the first years following the introduction it only appeared sporadically, but in the early 1980s there was an explosive increase in the lake-wide density of the Nile perch. At the same time there was a dramatic decrease in the abundance of cichlids and by the late 1980s they seldomly appeared in trawl catches (Fig. 6.16). Overfishing may have initiated the disappearance of cichlids, but introduction of the Nile perch is considered the main cause of the extinction. Cichlids also disappeared in areas without commercial fishing and gut content analysis of Nile perch showed that the major food item was cichlids until they had virtually disappeared. Cichlids are sensitive to heavy predation pressure because of their low reproductive potential; they produce a limited number of large eggs, which they brood in the mouth. In addition to predation, cichlids have suffered from the increasing eutrophication. One of the mechanisms explaining the speciose cichlid fauna is related to female mating preference for male colour patterns (Seehausen *et al.* 1997). Reproductive isolation between sympatric species has maintained this mating preference as other traits that may function as reproductive barriers have not evolved. However, the efficiency of the colour recognition is strongly dependent on light conditions and with the increasing turbidity with eutrophication the transmission properties for light has changed dramatically and thus affecting

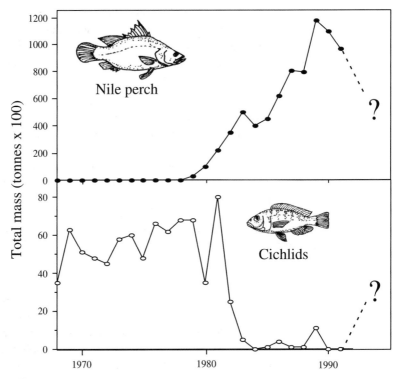

Fig. 6.16   Commercial landings of Nile perch (*Lates*) and cichlids in the Kenyan part of Lake Victoria since the late 1960s. Data from Gophen *et al*. (1995). Increasing fishing pressure during the 1990s have resulted in a decrease in Nile perch and an increase in cichlids populations (hatched lines).

transmission of colour signals. In experimental studies, Seehausen *et al.* (1997) showed that females showed no mating preference for colour patterns under poor light conditions. Thus, in turbid waters the reproductive barriers brake down, further contributing to the loss of species. Over 200 species (65%) have disappeared or are threatened by eradication. This mass extinction, which took place in less than a decade, may well be a tragic example of the largest extinction event in vertebrates during the last century.

Secondary effects of Nile perch introduction

The expansion of the Nile perch resulted in a three- to four-fold increase in commercial fish production which created much new employment. However, the intense fishing pressure in the 1990s have now resulted in a change in the size structure of Nile perch populations. Large individuals are gone and 70% of the catch now consists of immature fish, a typical sign of overexploited fish stocks. In the face of declining Nile perch populations, and thus reduced predation pressure, some of the cichlid species that were thought to have gone extinct have now reappeared. However, it is only a few species from a few trophic groups that now again appear at high densities. They represent species that are highly flexible with an ability to respond to environmental change such as eutrophication and changing mortality regimes due to fishery and predation.

Investigation of the Nile perch diet has shown that the juvenile Nile perch switched to feeding on a small, pelagic cyprinid (*Rastrineobola argenta*) and a freshwater prawn (*Caridina nilotica*) as the preferred food, fish, disappeared. Larger Nile perch have become cannibalistic. The detritivorous prawn and zooplanktivorous cyprinid increased dramatically when the cichlids were eradicated probably because of less competition for food by specialized detritivorous and zooplanktivorous cichlids, respectively. Anecdotal observations suggest that the Nile perch may have other indirect effects on the Lake Victoria ecosystem, such as an increase in phytoplankton, macrophytes, and many macroinvertebrate species. Large parts of the formerly oxygenated bottom waters are now also anoxic throughout the year, at times causing extensive fish kills.

In the short term, the introduction of Nile perch was a success story with respect to commercial fisheries, but recent indications of overfishing suggest that the Nile perch fisheries is not sustainable in the longer-term if the fishing pressure is kept at the present level. Implication of proper management strategies may create a sustainable fishery on Nile perch and it has been suggested (Balirwa *et al.* 2003) that if this fishery pressure is hard enough more 'extinct' cichlid species may reappear and, thus, allowing for a coexistence of Nile perch and native cichlids. However, as the reappearing cichlid species consists of a 'biologically filtered' subset (Fig. 6.2) of the original fish assemblage the resulting food web will probably be very different from the one that existed before human intervention. The introduction of Nile perch has caused irreversible damage to a highly complex freshwater system, of which we have probably not yet seen all the consequences. The possibility of understanding the evolution of the cichlid species flock and the mechanisms that allowed such a large number of morphologically similar species to coexist in Lake Victoria may have been lost forever. This ecological tragedy must, however, be seen in the perspective of availability of human food resources, and the risk of famine in this region.

Genetically modified organisms

A genetically modified organism (GMO) may also act as an exotic species in natural ecosystems. GMOs have a gene sequence that has been artificially modified through gene technology. The main reasons for modifying an organisms's gene sequence is to change its characteristics in a way that improves its human use, for example, by improving its resistance to diseases or increase its growth rate. Fish transgenic for growth hormone genes grow faster and become larger than non-transgenic fish (e.g. Hill *et al.* 2000). A major risk with transgenic fish seems to be that they escape from fish farms, interact with wild ancestors, and in this way become new actors in natural ecosystems (Reichhart 2000). For example, in many fish species larger males have a selective mating advantage and a higher reproductive success, suggesting that fast-growing transgenics may efficiently invade natural populations. Theoretical models have further suggested that if the high growth rate is combined with a viability disadvantage population fitness is reduced and eventually resulting in extinction of the natural population (Hedrick 2001).

The GMOs are one of the more recent potential environmental problems. Whether they will become a major environmental problem or not depends on policy- and decision-makers, as well as on the consumer market. The current global resistance to consumption of genetically modified crops such as soybeans and corn, may give a hint of a coming resistance to eating transgenic fish; a notion strengthened by the current economical problems the GMO industry is facing (Reichhart 2000).

Global differentiation

Much of the above, for example, effects of contaminants on lake and pond systems, has been based on patterns found in developed countries where industrialization caused many environmental problems, but where large efforts have been made to reduce the effects of at least point-source pollution. Developing countries, especially in the tropics, are facing problems at a completely different scale. Urbanization has resulted in strong pressure on freshwater systems for human consumption, while, simultaneously, these water sources are being increasingly polluted by untreated effluents from the same urban areas. At the same time resources harvested from freshwater systems, such as plants, invertebrates, and fish, constitute important parts of the diet in many developing countries. Water in these areas is thus, and will increasingly become, a limiting resource for human development and the importance of lakes and ponds as habitats for a diverse, natural biota will be secondary to the management of these systems for human needs.

Deforestation and increasing agricultural expansion directly affect lake and pond systems in developing countries through habitat destruction including landfilling, sedimentation, and water level reductions after irrigation, but indirect effects are also increasing in importance. The economy in developing countries is mainly based on agriculture and depends on intensive use of pesticides to improve productivity, creating a major problem of contamination of freshwater habitats (Lacher and Goldstein 1997). The use of pesticides in developing countries approaches that of, or is even higher than in, developed countries. At the same time legislation against organochlorine pesticides is less restrictive and safety measures to prohibit environmental effects are frequently not applied in tropical countries. Compounds that are banned in developed countries or new compounds not registered in their country of origin are exported for use in developing countries. Recent outbreaks of malaria have led to reintroduction of DDT in some areas. Further, although knowledge on the effects of contaminants in lakes in northern latitudes is incomplete, understanding of how they operate in tropical environments is negligible (Castillo *et al.* 1997, Lacher and Goldstein 1997). Freshwater habitats in the tropics differ from those in temperate zones in physical, chemical, and biological attributes and, for example, toxicity of contaminants may be higher in the tropics due to temperature effects on solubility, uptake, and bioconcentration.

The use of organochlorines in the tropics is not just of concern for countries in this region. Organochlorines are atmospherically transported from tropical regions, where pollutants are volatilized on the ground, towards the polar regions of the globe where the vapours are condensed and washed out. The 'condensation point' differs among organochlorine substances resulting in a global fractionation or distillation (Wania and Mackay 1993). DDT and PCB congeners, for example, which are considered 'semivolatile' are expected to deposit in temperate regions and potentially result in higher concentrations of these contaminants in the biota in the deposition than in the source areas (Larsson *et al.* 1995).

Increasing rates of waste disposal in lakes in developing countries will cause eutrophication problems. Industrialization in developing regions will also increase the emission of sulphur and although sulphur emissions are declining in Europe and North America (Fig. 6.11) they are rising in many developing countries around the world (Galloway 1995). Freshwater habitats in many areas of the world hitherto unaffected by acidic rain are at serious risk of acidification, as sulphur emissions exceed critical loads (Kuylenstierna *et al.* 2001). Further, as noted above, many of the examples of exotic species that have had the greatest impacts on lake and pond systems come from the tropics and there is no reason to believe that this will change in the future.

In an optimistic scenario, the developing countries will not make the same environmental mistakes as the developed countries, errors that are now costly to remedy. Policy- and decision-makers in developing countries may draw conclusions based on experiences from temperate systems when choosing among developmental pathways and thereby avoid errors made during the industrial era in Europe and North America. However, as pointed out above, tropical freshwater systems do not necessarily react to disturbances in the same way as temperate systems, and more research is needed. Economic aid from developed countries that is directed towards environmental conservation is also crucial for the future.

More pessimistically, it must be noted that in countries where the population increase results in a doubling of populations every 25–35 years, all aspects of the infrastructure need to increase at the same rate just to maintain present-day standards (Lacher and Goldstein 1997). Thus, the pressure on environmental values, especially water resources, will be tremendous and Lacher and Goldstein (1997) predict that environmental quality is doomed to decline in developing countries in foreseeable time.

Actions taken against environmental threats

Many environmental threats and pollutants have no county, federation, or national borders, but are spread globally by wind and water. Hence, one of the most important actions that can be taken against environmental problems is international agreements. Although such agreements may seem inefficient and slow, it may be the only way to reach a sustainable use of resources and an acceptable level of emissions on the international scale (Brönmark and Hansson 2002). The rapidly accumulating environmental problems have forced national, regional, and local administrations to improve the planning and protection of, for example, water resources. For example, the European Commission has passed the 'Water Framework Directive' (http://europa.eu.int/comm/environment/water/), which is a strategic plan agreed upon by all member states with the ultimate aim of completely stopping pollution of freshwater resources. The Directive is based upon two important strategies, namely that the management of water resources should be based on catchment areas, where all water resources should reach 'good' ecological quality. Furthermore, the principle that 'the polluter pays' should be applied, which means that if someone wants to use and pollute water resources, this user should pay for any damage caused to the common water resource. Several other important international agreements have been signed, including Agenda 21 (www.igc.org/habitat/agenda21/) and several initiatives from the United Nations. One such recent initiative regarding freshwater resources is the UNESCO World Water Assessment Programme (www.unesco.org/water/wwap) aimed at making the use of freshwaters more sustainable and reducing the number of people not having access to clean water. Although this chapter on 'Biodiversity and Environmental Threats' has given a relatively pessimistic view of the status of lake and pond ecosystems, we would like to end this book by stating that there are positive trends indicating that the work on environmental issues has been fruitful. Moreover, the growing concern for environmental problems, the implementation of new environmental strategies and administrations, as well as interest in international agreements, are indeed positive signs of changes that improve our possibilities to handle environmental threats in the future.

References

Acre, B. G. and Johnson, D. M. (1979). Switching and sigmoid functional response curves by damselfly naiads with alternative prey available. *Journal of Animal Ecology*, **48**, 703–20.

Adler, F. R. and Harvell, C. D. (1990). Inducible defenses, phenotypic variability and biotic environments. *Trends in Ecology and Evolution*, **5**, 407–10.

Andersson, G. (1984). The role of fish in lake ecosystems–and in limnology. *Norsk Limnologforening*, 189–97.

Andersson, G. and Cronberg, G. (1984). *Aphanizomenon flos-aquae* and large *Daphnia*–an interesting plankton association in hypertrophic waters. *Norsk Limnologforening*, 63–76.

Andersson, G., Granéli, W., and Stenson, J. (1988). The influence of animals on phosphorus cycling in lake ecosystems. *Hydrobiologia*, **170**, 267–84.

Appelberg, M. (1998). Restructuring of fish assemblages in Swedish lakes following amelioration of acid stress through liming. *Restoration Ecology*, **6**, 343–52.

Balirwa, J. S., Chapman, C. A., Chapman, L. J., Cowx, I. G., Geheb, K., Kaufman, L., Lowe-McConnel, R. H., Seehausen, O. Wanink, J. H. Welcomme, R. L., and Witte, F. (2003). Biodiversity and fishery sustainability in the lake Victoria basin: an unexpected marriage? *Bioscience*, **53**, 703–15.

Bärlocher, F., Mackay, R. J., and Wiggins, G. B. (1978). Detritus processing in a temporary vernal pond in southern Ontario. *Archiv für Hydrobiologie*, **81**, 269–95.

Begon, M., Harper, C. R., and Townsend, C. R. (1990). *Ecology. Individuals, populations and communities*, (2nd edn). Blackwell, Oxford.

Bergman, E. and Greenberg, L. A. (1994). Competition between a planktivore, a benthivore, and a species with ontogenetic diet shifts. *Ecology*, **75**, 1233–45.

Bially, A. and MacIsaac, H. J. (2000). Fouling mussels (*Dreissena* spp.) colonize soft sediments in lake Erie and facilitate benthic invertebrates. *Freshwater Biology*, **43**, 85–97.

Bilton, D. T., Freeland, J. R., and Okamura, B. (2001). Dispersal in freshwater invertebrates. *Annual Review of Ecology and Systematics*, **32**, 159–81.

Bird, D. F. and Kalff, J. (1986). Bacterial grazing by planktonic lake algae. *Science*, **231**, 493–5.

Bittner, K., Rothhaupt, K.-O., and Ebert, D. (2002). Ecological interactions of the microparasite *Caullerya mesnili* and its host *Daphnia galeata*. *Limnology and Oceanography*, **47**, 300–5.

Björk, S. (1988). Redevelopment of lake ecosystems–A case study report approach. *Ambio*, **17**, 90–8.

Blindow, I., Hargeby, A., and Andersson, G. (1998). Alternative stable states in shallow lakes–what causes a shift? In *The structuring role of submerged macrophytes in lakes*, (eds. E. Jeppesen, M. Søndergaard, and K. Christoffersen), pp. 353–68. Springer, Berlin.

Bloem, J., Albert, C., Bar-Gilissen, M.-J. G. *et al.* (1989*a*). Nutrient cycling through phytoplankton bacteria and protozoa in selectively filtered Lake Vechten water. *Journal of Plankton Research*, **11**, 119–31.

Bloem, J., Ellenbroek, F. M., Bär-Gilissen, M. J. B., and Cappenberg, T. E. (1989*b*). Protozoan grazing and bacterial production in stratified Lake Vechten estimated with fluorescently labeled bacteria and by thymidine incorporation. *Applied and Environmental Microbiology*, **55**, 1787–95.

Blomqvist, S., Gunnars, A., and Elmgren, R. (2004). Why limiting nutrients differ between temperate coastal seas and freshwater lakes: A matter of salt. *Limnology and Oceanography* (in press).

Boström, B., Jansson, M., and Forsberg, C. (1982). Phosphorus release from lake sediments. *Archiv für Hydrobiologie, Ergebnisse der Limnologie*, **18**, 5–59.

Bothwell, M., Sherbot, D., and Pollock, C. (1994). Ecosystem response to solar ultraviolet-B radiation: influence of trophic-level interactions. *Science*, **265**, 97–100.

Brönmark, C. (1985). Interactions between macrophytes, epiphytes and herbivores: an experimental approach. *Oikos*, **45**, 26–30.

Brönmark, C. (1994). Effects of tench and perch on interactions in a freshwater, benthic food chain. *Ecology*, **75**, 1818–24.

Brönmark, C. and Edenhamn, P. (1994). Does the presence of fish affect the distribution of tree frogs (*Hyla arborea*)? *Conservation Biology*, **8**, 841–5.

Brönmark, C. and Hansson, L.-A. (2002). Environmental issues in lakes and ponds: current state and future perspective. *Environmental Conservation*, **29**, 290–306.

Brönmark, C. and Miner, J. G. (1992). Predator-induced phenotypical change in crucian carp. *Science*, **258**, 1348–50.

Brönmark, C. and Pettersson, L. (1994). Chemical cues from piscivores induce a change in morphology in crucian carp. *Oikos*, **70**, 396–402.

Brönmark, C. and Weisner, S. E. B. (1992). Indirect effects of fish community structure on submerged vegetation in shallow, eutrophic lakes: an alternative mechanism. *Hydrobiologia*, **243/244**, 293–301.

Brönmark, C., Rundle, S. D., and Erlandsson, A. (1991). Interactions between freshwater snails and tadpoles: competition and facilitation. *Oecologia*, **87**, 8–18.

Brönmark, C., Klosiewski, S. P., and Stein, R. A. (1992). Indirect effects of predation in a freshwater, benthic food chain. *Ecology*, **73**, 1662–74.

Brooks, J. L. and Dodson, S. I. (1965). Predation, body size, and composition of plankton. *Science*, **150**, 28–35.

Browne, R. A. (1981). Lakes as islands: biogeographic distribution, turnover rates, and species composition in the lakes of central New York. *Journal of Biogeography*, **8**, 75–83.

Burns, C. (1968). The relationship between body size of filter-feeding cladocera and the maximum size of particle ingested. *Limnology and Oceanography*, **13**, 675–8.

Canter, H. M. (1979). Fungal and protozoan parasites and their importance in the ecology of phytoplankton. *Freshwater Biological Association* (Annual Report), **47**, 43–50.

Caraco, N., Cole, J. J., and Likens, G. E. (1989). Evidence for sulphate-controlled phosphorus release from sediments of aquatic systems. *Nature*, **341**, 316–18.

Carpenter, S. R. and Kitchell, J. F. (eds.) (1993). *The trophic cascade in lakes.* Cambridge University Press.

Carpenter, S. R., Kitchell, J. F., and Hodgson, J. R. (1985). Cascading trophic interactions and lake productivity: fish predation and herbivory can regulate lake ecosystems. *Bioscience*, **35**, 634–9.

Castillo, L. E., de la Cruz, E., and Ruepert, C. (1997). Ecotoxicology and pesticides in tropical aquatic ecosystems of Central America. *Environmental Toxicology and Chemistry*, **16**, 41–51.

Chambers, P. A. and Kalff, J. (1985). Depth distribution and biomass of submersed aquatic macrophyte communities in relation to Secchi depth. *Canadian Journal of Fisheries and Aquatic Sciences*, **42**, 701–9.

Chen, C. Y. and Folt, C. L. (1996). Consequences of fall warming for zooplankton overwintering success. *Limnology and Oceanography*, **41**, 1077–86.

Chen, C. Y., Stemberger, R. S., Klaue, B., Blum, J. D., Pickhardt, P. C., and Folt, C. L. (2000). Accumulation of heavy metals in food web components across gradients of lakes. *Limnology and Oceanography*, **45**, 1525–36.

Cohen, G. M. and Shurin, J. B. (2003) Scale-dependence and mechanisms of dispersal in freshwater zooplankton. *Oikos*, **103**, 603–17.

Cole, J. J., Pace, M. L., Carpenter, S. R., and Kitchell, J. F. (2000). Persistence of net heterotrophy in lakes during nutrient addition and food web manipulations. *Limnology and Oceanography*, **45**, 1718–30.

Connell, J. H. (1978). Diversity in tropical rain forests and coral reefs. *Science*, **199**, 1302–10.

Craig, J. F. (1987). *The biology of perch and related fish.* Croom Helm, Beckenham, UK.

Cronberg, G. (1982). *Pediastrum* and *Scenedesmus* (Chlorococcales) in sediments from Lake Växjösjön, Sweden. *Archiv für Hydrobiologie*, Supplement, **60**, 500–7.

Cuker, B. (1983). Competition and coexistence among the grazing snail *Lymnea*, Chironomidae, and microcrustacea in an arctic lacustrine community. *Ecology*, **64**, 10–15.

Cummins, K. (1973). Trophic relations of aquatic insects. *Annual Review of Entomology*, **18**, 183–206.

Dacey, J. W. H. (1981). Pressurized ventilation in the yellow waterlily. *Ecology*, **62**, 1137–47.

Dawidowicz, P. and Loose, C. J. (1992). Metabolic costs during predator-induced diel vertical migration in *Daphnia*. *Limnology and Oceanography*, **37**, 1589–95.

Dawidowicz, P., Pijanowska, J., and Ciechomski, K. (1990). Vertical migration of *Chaoborus* larvae is induced by the presence of fish. *Limnology and Oceanography*, **35**, 1631–7.

Dawkins, R. and Krebs, J. R. (1979). Arms races between and within species. *Proceedings of the Royal Society of London, Series B*, **205**, 489–511.

Diehl, S. (1988). Foraging efficiency of three freshwater fishes: effects of structural complexity and light. *Oikos*, **53**, 207–14.

Diehl, S. (1992). Fish predation and benthic community structure: the role of omnivory and habitat complexity. *Ecology*, **73**, 1646–61.

Diehl, S. (1993). Relative consumer sizes and the strengths of direct and indirect interactions in omnivorous feeding relationships. *Oikos*, **68**, 151–7.

Dodson, S. I., Arnott, S. E., and Cottingham, K. L. (2000). The relationship in lake communities between primary productivity and species richness. *Ecology*, **81**, 2662–79.

Driscoll, C. T., Lawrence, G. B., Bulger, A. T., Butler, T. J., Cronan, C. S., Eagar, C., Lambert, K. F., Likens, G. E., Stoddard, J. L., and Weathers, K. C. (2001). Acidic deposition in the northeastern United States: sources and inputs, ecosystem effects, and management strategies. *Bioscience*, **51**, 180–98.

Dvorac, J. and Best, E. P. H. (1982). Macroinvertebrate communities associated with the macrophytes of Lake Vechten: structural and functional relationships. *Hydrobiologia*, **95**, 115–26.

Ebenman, B. and Persson, L. (eds.) (1988). *Size-structured populations. Ecology and evolution.* Springer, Berlin.

Edmunds, M. (1974). *Defence in animals.* Longman, Essex, UK.

Ehlinger, T. J. (1990). Habitat choice and phenotype-limited feeding efficiency in bluegill: individual differences and trophic polymorphism. *Ecology*, **71**, 886–96.

Eklöv, P. and Diehl, S. (1994). Piscivore efficiency and refuging prey: the importance of predator search mode. *Oecologia*, **98**, 344–53.

Eklöv, P. and VanKooten, T. (2001). Facilitation among piscivorous predators: effects of prey habitat use. *Ecology*, **82**, 2486–94.

Elliot, J. M. (1981). Some aspects of thermal stress on freshwater teleosts. In *Stress and fish*, (ed. A. D. Pickering, pp. 209–45). Academic Press, London.

Elser, J. J. and Hassett, P. (1994). A stoichiometric analysis of the zooplankton–phytoplankton interaction in marine and freshwater ecosystems. *Nature*, **370**, 211–13.

Engelhardt, K. A. M. and Ritchie, M. E. (2001). Effects of macrophyte species richness on wetland ecosystem functioning and services. *Nature*, **411**, 687–9.

Ewald, G., Larsson, P., Linge, H., Okla, L., and Szarzi, N. (1998). Biotransport of organic pollutants to an inland Alaska lake by migrating sockeye salmon (*Oncorhyncus nerka*). *Arctic*, **51**, 40–7.

Fauth, J. E. (1990). Interactive effects of predators and early larval dynamics of the treefrog *Hyla chrysoscelis*. *Ecology*, **71**, 1609–16.

Fitzsimons, J. D., Leach, J. H., Nepszy, S. J., and Cairns, V. W. (1995). Impacts of zebra mussel on walleye (*Stizstedion vitreum*) reproduction in western Lake Erie. *Canadian Journal of Fisheries and Aquatic Sciences*, **52**, 578–86.

Flöder, S. and Sommer, U. (1999). Diversity in planktonic communities: an experimental test of the intermediate disturbance hypothesis. *Limnology and Oceanography*, **44**, 1114–19.

Forbes, S. (1925). The lake as a microcosm. *Bulletin of the Illinois Natural History Survey*, **15**, 537–50. (Originally published 1887.)

Fox, L. R. and Murdoch, W. W. (1978). Effects of feeding history on short-term and long-term functional responses in *Notonecta hoffmanni*. *Journal of Ecology*, **47**, 945–59.

Frantz, T. C. and Cordone, A. J. (1967). Observations on deepwater plants in Lake Tahoe, California and Nevada. *Ecology*, **48**, 709–14.

Fretwell, S. D. (1977). The regulation of plant communities by the food chains exploiting them. *Perspectives in Medicine and Biology*, **20**, 169–85.

Frost, T. M. and Williamson, C. E. (1980). In situ determination of the effect of symbiotic algae on the growth of the freshwater sponge *Spongilla lacustris*. *Ecology*, **61**, 1361–70.

Frost, T. M., de Nagy, G. S., and Gilbert, J. J. (1982). Population dynamics and standing biomass of the freshwater sponge, *Spongilla lacustris*. *Ecology*, **63**, 1203–10.

Fry, B. (1991). Stable isotope diagrams of freshwater food webs. *Ecology*, **72**, 2293–7.

Fuhrman, J. A. (1999). Marine viruses and their biogeochemical and ecological effects. *Nature*, **399**, 541–8.

Galloway, J. (1995). Acid deposition: perspectives in time and space. *Water, Air and Soil Pollution*, **85**, 15–24.

Gause, G. F. (1934). *The struggle for existence*. Williams & Wilkins, Baltimore, MD.

George, D. G. (1981). Zooplankton patchiness. *Report of the Freshwater Biological Association*, **49**, 32–44.

Gilinsky, E. (1984). The role of fish predation and spatial heterogeneity in determining benthic community structure. *Ecology*, **65**, 455–68.

Giller, P. S. (1984). *Community structure and the niche*. Chapman & Hall, London.

Goldstein, S. F. (1992). Flagellar beat patterns in algae. In *Algal cell motility*, (ed. M. Melkonian), *Current phycology*, Vol. 3, pp. 99–153. Chapman & Hall, New York.

Gophen, M., Ochumba, P. B. O., and Kaufman, L. S. (1995). Some aspects of perturbations in the structure and biodiversity of the ecosystem of Lake Victoria (East Africa). *Aquatic Living Resources*, **8**, 27–41.

Grace, J. B. and Wetzel, R. G. (1981). Habitat partitioning and competitive displacement in cattails (*Typha*): experimental field studies. *American Naturalist*, **118**, 463–74.

Graham, J. B. (1988). Ecological and evolutionary aspects of integumentary respiration: body size, diffusion and the invertebrata. *American Zoologist*, **28**, 1031–45.

Graham, J. B. (1990). Ecological, evolutionary and physical factors influencing aquatic animal respiration. *American Zoologist*, **30**, 137–46.

Grey, J., Jones, R. I., and Sleep, D. (2001). Seasonal changes in the importance of the source of organic matter to the diet of zooplankton in Loch Ness, as indicated by stable isotope analysis. *Limnology and Oceanography*, **46**, 505–13.

Gunn, J. and Keller, W. (1990). Biological recovery of an acidified lake after reductions in industrial emissions of sulphur. *Nature*, **345**, 431–3.

Gustafsson, S. and Hansson, L.-A. (2004). Development of tolerance against toxic cyanobacteria in *Daphnia*. *Aquatic Ecology*, **38**, 37–44.

Gyllström, M. and Hansson, L.-A. (2004). Dormancy in freshwater zooplankton: induction, termination and the importance of benthic–pelagic coupling. *Aquatic Sciences*, **66**, 1–22.

Hairston, N. G., Jr. (1987). Diapause as a predator-avoidance adaptation. In *Predation. Direct and indirect impacts in aquatic communities*, (eds. W. C. Kerfoot and A. Sih), pp. 281–90. University Press of New England, Hanover, NH.

Hairston, N. G., Smith, F. E., and Slobodkin, L. B. (1960). Community structure, population control, and competition. *American Naturalist*, **94**, 421–5.

Hairston, N. G., Jr. van Brunt, R., Kearns, C., and Engstrom, D. R. (1995). Age and survivorship of diapausing eggs in a sediment egg bank. *Ecology*, **76**, 1706–11.

Hairston, N. G., Jr. Lampert, W., Cáceres, C. E., Holtmeier, C. L., Weider, L. J., Gaedke, U., Fischer, J. M., Fox, J. A., and Post, D. M. (1999). Rapid evolution revealed by dormant eggs. *Nature*, **401**, 446.

Hambright, K. D. (1994). Morphological constraints in the piscivore–planktivore interaction: implications for the trophic cascade hypothesis. *Limnology and Oceanography*, **39**, 897–912.

Hambright, K. D., Drenner, R. W., McComas, S. R., and Hairston, N. G., Jr. (1991). Gape-limited piscivores, planktivore size refuges, and the trophic cascade hypothesis. *Hydrobiologia*, **121**, 389–404.

Hansson, L.-A. (1992a). Factors regulating periphytic algal biomass. *Limnology and Oceanography*, **37**, 322–8.

Hansson, L.-A. (1992b). The role of food chain composition and nutrient availability in shaping algal biomass development. *Ecology*, **73**, 241–7.

Hansson, L.-A. (1996). Behavioural response in plants: adjustment in algal recruitment induced by herbivores. *Proceedings of the Royal Society of London, Series B*, **263**, 1241–4.

Hansson, L.-A. (2000). Synergistic effects of food web dynamics and induced behavioral responses in aquatic ecosystems. *Ecology*, **81**, 842–51.

Hansson, L.-A. (2004). Phenotypic plasticity in pigmentation among zooplankton induced by conflicting threats from predation and UV radiation. *Ecology*, **85**, 1005–16.

Hansson, L.-A., Bergman, E., and Cronberg, G. (1998a). Size structure and succession in phytoplankton communities: the impact of interactions between herbivory and predation. *Oikos*, **81**, 337–45.

Hansson, L.-A., Annadotter, H., Bergman, E., Hamrin, S.F., Jeppessen, E. Kairesalo, T., Luokkanen, E., Nilsson, P.-Å., Søndergaard, Ma., & Strand, J. (1998b). Biomanipulation as an application of food chain theory: constraints, synthesis and recommondations for temperate lakes *Ecosystems*, **1**, 558–74.

Hargeby, A., Blindow, I., and Hansson, L.-A. (2004). Shifts between clear and turbid states in a shallow lake: multi-causal stress from climate, nutrients and biotic interactions. *Archiv für Hydrobiologie*, (in press).

Havel, J. E. (1987). Predator-induced defenses: a review. In *Predation. Direct and indirect impacts in aquatic communities*, (eds. W. C. Kerrfoot and A. Sih), pp. 263–78. University Press of New England, Hanover, NH.

Hedrick, P. W. (2001). Invasion of transgenes from salmon or other genetically modified organisms into natural populations. *Canadian Journal of Fisheries and Aquatic Science*, **58**, 841–4.

Hessen, D. O. and van Donk, E. (1993). Morphological changes in *Scenedesmus* induced by substances released from *Daphnia*. *Archiv für Hydrobiologie*, **127**, 129–40.

Hill, H. Z. (1992). The function of melanin or six blind people examine an elephant. *BioEssays*, **14**, 49–56.

Hill, J. A., Kiessling, A., and Devlin, R. H. (2000). Coho salmon (*Oncorhyncus kisutch*) transgenic for a growth hormone gene construct exhibit increased rates of muscle hyperplasia and detectable levels of differential gene expression. *Canadian Journal of Fisheries and Aquatic Sciences*, **57**, 939–50.

Hillebrand, H. and Azovsky, A. I. (2001). Body size determines the strength of the latitudinal diversity gradient. *Ecography*, **24**, 251–6.

Holling, C. S. (1959). Some characteristics of simple types of predation and parasitism. *Canadian Entomologist*, **91**, 385–98.

Hoogland, R., Morris, D., and Tinbergen, N. (1957). The spines of sticklebacks (*Gasterosteus* and *Pygosteus*) as means of defence against predators (*Perca* and *Esox*). *Behaviour*, **10**, 205–36.

Houlahan, J. E., Findlay, C. S., Schmidt, B. R., Meyer, A. H., and Kuzmin, S. L. (2000). Quantitative evidence for global amphibian population decline. *Nature*, **404**, 752–5.

Hrbácek, J., Dvorakova, M., Korínek, V., and Prochákóva, L. (1961). Demonstration of the effect of the fish stock on the species composition of zooplankton and the intensity of metabolism of the whole plankton association. *Verhandlungen Internationale Vereinigung für Theoretische and Angewandte Limnologie*, **14**, 192–5.

Hutchinson, G. E. (1957). Concluding remarks. *Cold Spring Harbor Symposium on Quantitative Biology*, **22**, 415–27.

Hutchinson, G. E. (1961). The paradox of the plankton. *American Naturalist*, **95**, 137–45.

Hutchinson, G. E. (1967). *A treatise on limnology*. Vol. II. *Introduction to lake biology and the limnoplankton*. Wiley, New York.

Hutchinson, G. E. (1993). *A treatise on limnology*. Vol. IV. *The zoobenthos*. Wiley, New York.

Jansson, M., Blomqvist, P., Jonsson, A., and Bergström, A.-K. (1996). Nutrient limitation of bacterioplankton, autotrophic and mixotrophic phytoplankton, and heterotrophic nanoflagellates in lake Örträsket. *Limnology and Oceanography*, **41**, 1552–9.

Jeffries, D. S., Clair, T. A., Couture, S., Dillon, P. J., Dupont, J., Keller, W., McNicol, D. K., Turner, M. A., Vet, R., and Weeber, R. (2003). Assessing the recovery of lakes in southeastern Canada from the effects of acidic deposition. *Ambio*, **32**, 176–82.

Jeffries, M. J. and Lawton, J. H. (1984). Enemy free space and the structure of ecological communities. *Biological Journal of the Linnean Society*, **23**, 269–86.

Jenkins, M. (2003). Prospects for biodiversity. *Science*, **302**, 1175–7.

Jeppesen, E., Jensen, J., Søndergaard, M., Lauridsen, T., Junge Pedersen, L., and Jensen, L. (1996). Top–down control in freshwater lakes: the role of fish, submerged macrophytes and water depth. *Hydrobiologia*, **342/343**, 151–64.

Jeppesen, E., Søndergaard, M. A., Søndergaard, M. O. , and Christoffersen, K. (eds.) (1998). *The structuring role of submerged macrophytes in lakes*. Springer. NewYork.

Jeppesen, E., Leavitt, P., De Meester, L., and Jensen, J. P. (2001). Functional ecology and paleolimnology: using cladoceran remains to reconstruct anthropogenic impact. *Trends in Ecology and Evolution*, **16**, 191–8.

Johansson, F. (1992). Effects of zooplankton availability and foraging mode on cannibalism in three dragonfly larvae. *Oecologia*, **91**, 179–83.

Johnson, M. and Malmqvist, B. (2000). Ecosystem process rate increases with animal species richness: evidence from leaf-eating, aquatic insects. *Oikos*, **89**, 519–23.

Johnson, N., Revenga, C., and Echeverria, J. (2001). Managing water for people and nature. *Science*, **292**, 1071–2.

Jones, J. I. and Sayer, C. D. (2003). Does the fish–invertebrate–periphyton cascade precipitate plant loss in shallow lakes? *Ecology*, **84**, 2155–67.

Jones, R. I. (2000). Mixotrophy in planktonic protists: an overview. *Freshwater Biology*, **45**, 219–26.

Jürgens, K. and Güde, H. (1994). The potential importance of grazing-resistant bacteria in planktonic systems. *Marine Ecology Progress Series*, **112**, 169–88.

Kaufman, L. (1992). Catastrophic change in species-rich freshwater ecosystems. *Bioscience*, **42**, 846–58.

Kerfoot, W. C. (1987). Cascading effects and indirect pathways. In *Predation. Direct and indirect impacts in aquatic communities* (eds. W. C. Kerfoot and A. Sih), pp. 57–70. University Press of New England, Hanover, NH.

Kerfoot, W. C. and Sih, A. (eds.) (1987). *Predation. Direct and indirect impacts on aquatic communities*. University Press of New England, Hanover, NH.

Kiesecker, J., Blaustein, A., and Belden, L. (2001). Complex causes of amphibian population declines. *Nature*, **410**, 681–3.

Kiesecker, J. M. and Skelly, D. K. (2000). Choice of oviposition site by gray treefrogs: the role of potential parasitic infection. *Ecology*, **81**, 2939–43.

Kinzig, A. P., Pacala, S. W., and Tilman, D. (eds.) (2001). The functional consequences of biodiversity. Princeton University Press, Princeton and Oxford.

Kling, G. W., Hayhoe, K., Johnson, L. B., Magnuson, J. J., Polasky, S., Robinson, S. K., Shuter, B. J., Wander, M. M., Wuebbles, D. J., Zak, D. R., Lindroth, R. L., Moser, S. C., and Wilson, M. L. (2003). Confronting climate change in the Great Lakes region: impacts on communities and ecosystems. Union of Concerned Scientists, Cambridge, MA, and Ecological Society of America, Washington, DC (www.ucsusa.org/greatlakes/).

Kolar, C. S. and Lodge, D. M. (2000). Freshwater nonindigenous species: interactions with other global changes. In *Invasive species in a changing world*, (eds. H. A. Mooney and R. J. Hobbs), pp. 3–30. Island Press, Washington, DC.

Krebs, C. J. (2001). *Ecology*, (5th edn.). Benjamin Cummings, San Francisco, USA.

Kuylenstierna, J., Rodhe, H., Cinderby, S., and Hicks, K. (2001). Acidification in developing countries: ecosystem sensitivity and the critical load approach on a global scale. *Ambio*, **30**, 20–8.

Lacher, T. E. and Goldstein, M. I. (1997). Tropical ecotoxicology: status and needs. *Environmental Toxicology and Chemistry*, **16**, 100–11.

Lampert, W. (1993). Ultimate causes of diel vertical migration of zooplankton: new evidence for the predator-avoidance hypothesis. *Archiv für Hydrobiologie Beiheft Ergebnisse der Limnologie*, **39**, 79–88.

Lampert, W. (1994). Phenotypic plasticity of the filter screens in *Daphnia*: adaptation to a low food environment. *Limnology and Oceanography*, **39**, 997–1006.

Lampert, W., Rothhaupt, K. O., and von Elert, E. (1994). Chemical induction of colony formation in a green alga (*Scenedesmus acutus*) by grazers (*Daphnia*). *Limnology and Oceanography*, **39**, 1543–50.

Langeland, A., L'Abée-Lund, J. H., Jonsson, B., and Jonsson, N. (1991). Resource partitioning and niche shift in arctic charr *Salvelinus alpinus* and brown trout *Salmo trutta*. *Journal of Animal Ecology*, **60**, 895–912.

Larsson, P., Berglund, O., Backe, C., Bremle, G., Eklöv, A., Järnmark, C., and Persson, A. (1995). DDT – fate in tropical and temperate regions. *Naturwissenschaften*, **82**, 559–61.

Lauridsen, T. L. and Buenk, I. (1996). Diel changes in the horizontal distribution of zooplankton in the littoral zone of two shallow eutrophic lakes. *Archiv für Hydrobiologie*, **137**, 161–76.

Lauridsen, T. L. and Lodge, D. M. (1996). Avoidance by *Daphnia magna* of fish and macrophytes: chemical cues and predator-mediated use of macrophyte habitat. *Limnology and Oceanography*, **41**, 794–8.

Laurila, A., Kujasalo, J., and Ranta, E. (1997). Different antipredator behaviour in two anuran tadpoles: effects of predator diet. *Behavioural Ecology and Sociobiology*, **40**, 329–36.

Lawton, J. H. (1994). What do species do in ecosystems? *Oikos*, **71**, 367–374.

Lean, D. R. S. (1973). Phosphorus dynamics in lake water. *Science*, **179**, 678–80.

Lehman, J. T. and Sandgren, C. D. (1985). Species-specific rates of growth and grazing loss among freshwater algae. *Limnology and Oceanography*, **30**, 34–46.

Lindell, M. J., Granéli, W., and Tranvik, L. J. (1995). Enhanced bacterial growth in response to photochemical transformation of dissolved organic matter. *Limnology and Oceanography*, **40**, 195–9.

Lodge, D. M. (1993). Biological invasions: lessons for ecology. *Trends in Ecology and Systematics*, **8**, 133–7.

Lodge, D. M., Brown, K. M., Klosiewski, S. P., Stein, R. A., Covich, A. P., Leathers, B. K., and Brönmark, C. (1987). Distribution of freshwater snails: spatial scale and the relative importance of physiochemical and biotic factors. *American Malacological Bulletin*, **5**, 73–84.

Lodge, D. M., Barko, J. W., Strayer, D., Melack, J. M., Mittelbach, G. G., Howarth, R.W. *et al.* (1988). Spatial heterogeneity and habitat interactions in lake communities. In *Complex interactions in lake communities*, (ed. S. R. Carpenter), pp. 181–209. Springer, New York.

Lodge, D. M., Kershner, M. W., and Aloi, J. (1994). Effects of an omnivorous crayfish (*Orconectes rusticus*) on a freshwater littoral food web. *Ecology*, **75**, 1265–81.

Loganathan, B. G. and Kannan, K. (1994). Global organochlorine contamination trends: an overview. *Ambio*, **23**, 187–91.

Loot, G., Aulagnier, S., Lek, S., Thomas, F., and Guégan, J.-F. (2002). Experimental demonstration of a behavioural modification in a cyprinid fish, *Rutilus rutilus* (L.), induced by a parasite, *Ligula intestinalis* (L.). *Canadian Journal of Zoology*, **80**, 738–44.

Loreau, M., Naeem, S., and Inchausti, P. (2002). *Biodiversity and ecosystem functioning*. Oxford University Press.

Ludyanskiy, M. L., McDonald, D., and MacNeill, D. (1993). Impact of the zebra mussel, a bivalve invader. *Bioscience*, **43**, 533–44.

Luecke, C. and O'Brien, W. J. (1981). Phototoxicity and fish predation: selective factors in color morphs in *Heterocope*. *Limnology and Oceanography*, **26**, 454–60.

Lürling, M. and van Donk, E. (1997). Life history consequences for *Daphnia pulex* feeding on nutrient-limited phytoplankton. *Freshwater Biology*, **38**, 639–709.

Maberly, S. C., King, L., Dent, M. M., Jones, R. I., and Gibson, C. E. (2002). Nutrient limitation of phytoplankton and periphyton growth in upland lakes. *Freshwater Biology*, **47**, 2136–52.

MacArthur, R. H. (1955). Fluctuations of animal populations and a measure of community stability. *Ecology*, **36**, 533–6.

MacArthur, R. H. and Wilson, E. O. (1967). *The theory of island biogeography*. Princeton University Press, Princeton, NJ.

Madsen, J. D. and Adams, M. S. (1989). The light and temperature dependence of photosynthesis and respiration in *Potamogeton pectinatus*. *Aquatic Botany*, **36**, 23–31.

Magnuson, J. J., Crowder, L. B., and Medvick, P. A. (1979). Temperature as an ecological resource. *American Zoologist*, **19**, 331–43.

Magnuson, J. J., Beckel, A., Mills, K., and Brandt, S. B. (1985). Surviving winter hypoxia: behavioral adaptations of fishes in a northern Wisconsin winterkill lake. *Environmental Biology of Fishes*, **14**, 241–50.

Magnusson, J. J., Webster, K. E., Assell, R. A., Browser, C. J., Dillon, P. J., Eaton, J. G., Evans, H. E., Fee, E. J., Hall, R. I., Mortsch, L. R., Schindler, D. W. and Quinn, F. H. (1997). Potential effects of climate changes on aquatic systems: Laurentian Great Lakes and precambrian shield region. *Hydrological Processes*, **11**, 825–71.

Malinen, T., Horppila, J., and Liljendahl-Nurminen, A. (2001). Langmuir circulations disturb the low-oxygen refuge of phantom midge larvae. *Limnology and Oceanography*, **46**, 689–92.

Martin, P. and Lefebvre, M. (1995). Malaria and climate: sensitivity of malaria potential transmission to climate. *Ambio*, **24**, 200–7.

Martin, T. H., Crowder, L. B., Dumas, C. F., and Burkholder, J. M. (1994). Indirect effects of fish on macrophytes in Bays Mountain Lake: evidence for a littoral trophic cascade. *Oecologia*, **89**, 476–81.

Mathiessen, P. (2000). Is endocrine disruption a significant ecological issue? *Ecotoxicology*, **9**, 21–4.

Mathiessen, P. and Sumpter, J. P. (1998). Effects of estrogenic substances in the aquatic environment. In *Fish ecotoxicology*, (eds. T. Braunbeck, D. Hinton, and B. Streit), pp. 319–335. Birkhauser Verlag. Basel, Switzerland.

May, R. (1977). Thresholds and breakpoints in ecosystems with a multiplicity of stable states. *Nature*, **269**, 471–7.

Mazumder, A. and Taylor, W. D. (1994). Thermal structure of lakes varying in size and water clarity. *Limnology and Oceanography*, **39**, 968–76.

Mazumder, A., Taylor, W., McQueen, D., and Lean, D. (1990). Effects of fish and plankton on lake temperature and mixing depth. *Science*, **247**, 312–15.

McCollum, S. A. and Leimberger, J. D. (1997). Predator-induced morphological changes in an amphibian: predation by dragonflies affect tadpole shape and color. *Oecologia*, **109**, 615–21.

McGrady-Steed, J., Harris, P. M., and Morin, P. J. (1997). Biodiversity regulates ecosystem predictability. *Nature*, **390**, 162–5.

McPeek, M. A. (1998). The consequences of changing the top predator in a food web: a comparative experimental approach. *Ecological Monographs*, **68**, 1–23.

McQueen, D. J., Post, J. R., and Mills, E. L. (1986). Trophic relationships in freshwater pelagic ecosystems. *Canadian Journal of Fisheries and Aquatic Sciences*, **43**, 1571–81.

Mermillod-Blondin, F., Gerino, M., Creuze des Chatelliers, M., and Degrange, V. (2002). Functional diversity among three detritivorous hyporheic invertebrates: an experimental study in microcosms. *Journal of the North American Benthological Society*, **21**, 132–49.

Meyer, A. (1987). Phenotypic plasticity and heterochrony in *Cichlasoma managuense* (Pisces, Cichlidae) and their implications for speciation in cichlids fishes. *Evolution*, **41**, 1357–69.

Middelboe, A. L. and Markager, S. (1997) Depth limits and minimum light requirements of freshwater macrophytes. *Freshwater Biology*, **37**, 553–68.

Milinski, M. (1985). Risk of predation of parasitized sticklebacks (*Gasterosteus aculeatus* L.) under competition for food. *Behaviour*, **93**, 203–16.

Milinski, M. and Heller, R. (1978). Influence of a predator on the optimal foraging behaviour of sticklebacks (*Gasterosteus aculeatus* L.). *Nature*, 275, 642–4.

Mills, E. L., Leach, J. H., Carlton, J. T., and Secor, C. L. (1994). Exotic species and the integrity of the Great Lakes. *Bioscience*, **44**, 666–75.

Mittelbach, G. G. (1988). Competition among refuging sunfishes and effects of fish density on littoral zone invertebrates. *Ecology*, **69**, 614–23.

Mittelbach, G. G. and Chesson, P. L. (1987). Predation risk: indirect effects on fish populations. In *Predation: direct and indirect effects*, (eds. W. C. Kefoot and A. Sih), pp. 315–22. University Press of New England, Hanover, NH.

Mittelbach, G. G. and Osenberg, C. W. (1993). Stage-structured interactions in bluegill: consequences of adult resource variation. *Ecology*, **74**, 2381–94.

Mittelbach, G. G., Osenberg, C. W., and Leibold, M. A. (1988). Trophic relations and ontogenetic niche shifts in aquatic ecosystems. In *Size-structured populations. Ecology and evolution*, (eds. B. Ebenman and L. Persson), pp. 219–35. Springer, Berlin.

Mittelbach, G. G., Steiner, C. F., Scheiner, S. M., Gross, K. L., Reynolds, H. L., Waide, R. B., Willig, M. R., Dodson, S. I., and Gough, L. (2001). What is the observed relationship between species richness and productivity? *Ecology*, **82**, 2381–96.

Moss, B. (1980). *Ecology of fresh waters*. Blackwell, Oxford.

Moss, B., Jones, P., and Phillips, G. (1994). August Thienemann and Loch Lomond – an approach to the design of a system for monitoring the state of north-temperate standing waters. *Hydrobiologia*, **290**, 1–12.

Moss, B., Stansfield, J., Irvine, K., Perrow, M., and Phillips, G. (1996). Progressive restoration of a shallow lake: a 12-year experiment on isolation, sediment removal and biomanipulation. *Journal of Applied Ecology*, **33**, 71–86.

Moss, B., Stephen, D., Alvarez, C., Becares, E., van de Bund, W., van Donk, E., de Eyto, E., Feldmann, T., Fernández-Aláez, C., Fernández-Aláez, M., Franken, R. J. M., García-Criado F., Gross, E., Gyllström, M., Hansson, L-A., Irvine, K., Järvalt, A., Jenssen, J-P., Jeppesen, E., Kairesalo, T., Kornijow, R., Krause, T., Künnap, H., Laas, A., Lill, E., Luup, H., Miracle, M. R., Nõges, P., Nõges, T., Nykannen, M., Ott, I., Peeters, E.T.H.M., Phillips, G., Romo, S., Salujõe, J., Scheffer, M., Siewertsen, K., Tesch, C., Timm, H., Tuvikene, L., Tonno, I., Vakilainnen, K., and Virro, T. (2002). The determination of ecological quality in shallow lakes – a tested expert system (ECOFRAME) for implementation of the European Water Framework Directive. *Aquatic Conservation*, **13**, 507–49.

Muir, D. C. G., Ford, C. A., Grift, N. P., Metner, D. A., and Lockhart, W. L. (1990). Geographic variation in chlorinated hydrocarbons in burbot (*Lota lota*) from remote lakes and rivers in Canada. *Archives of Environmental Contamination and Toxicology*, **5**, 29–40.

Murdoch, W. W., Scott, M. A., and Ebsworth, P. (1984). Effects of a general predator, *Notonecta* (Hemiptera), upon a freshwater community. *Journal of Animal Ecology*, **53**, 791–808.

Naeem, S. and Li, S. (1998). Consumer species richness and autotrophic biomass. *Ecology*, **79**, 2603–15.

Neill, W. E. (1990). Induced vertical migration in copepods as a defence against invertebrate predation. *Nature*, **345**, 524–6.

Neverman, D. and Wurtsbaugh, W. A. (1994). The thermoregulatory function of diel vertical migration for a juvenile fish, *Cottus extensus*. *Oecologia*, **98**, 247–56.

Nilsson, N.-A. (1965). Food segregation between salmonid species in north Sweden. *Reports from the Institute of Freshwater Research, Drottningholm*, **46**, 58–78.

Nyström, P., Brönmark, C., and Granéli, W. (1999). Influence of an exotic and a native crayfish species on a littoral benthic community. *Oikos*, **85**, 545–53.

Oksanen, L., Fretwell, S. D., Arruda, J., and Niemela, P. (1981). Exploitation eco-systems in gradients of primary productivity. *American Naturalist*, **118**, 240–61.

Olson, M. H., Mittelbach, G. G., and Osenberg, C. W. (1995). Competition between predator and prey: resource-based mechanisms and implications for stage-structured dynamics. *Ecology*, **76**, 1758–71.

O'Reilly, G. M., Allin, S. R., Pilsnier, P.-D., Cohen, A. S., and Mckee, B. (2003). Climate change decreases aquatic ecosystem productivity of Lake Tanganyika, Africa. *Nature*, **424**, 766–8.

Osenberg, C. W., Mittelbach, G. G., and Wainwright, P. C. (1992). Two-stage life histories in fish: the interaction between juvenile competition and adult - performance. *Ecology*, **73**, 255–67.

Pace, M. L., Cole, J. J. Carpenter, S. R., Kitchell, J. F., Hodgson, J. R., Van de Bogert, M. C., Blake, D. L., Kritzberg, E. S., and Bastviken, D. (2004). Whole-lake carbon-13 additions reveal terrestrial support of aquatic food webs. *Nature*, **427**, 240–3.

Parker, I. M., Simberloff, D., Lonsdale, W. M., Goodell, K., Wonham, M., Kareiva, P. M., Williamson, M. H., Von Holle, B., Moyle, P. B., Byers, J. E., and Goldwasser, L. (1999). Impact: toward a framework for understanding the ecological effects of invaders. *Biological Invasions*, **1**, 3–19.

Peckarsky, B. L. (1984). Predator–prey interactions among aquatic insects. In *The ecology of aquatic insects*, (eds. V. H. Resh and D. M. Rosenberg), pp. 196–254. Praeger, New York.

Persson, L. (1988). Asymmetries in competitive and predatory interactions in fish populations. In *Size-structured populations. Ecology and evolution* (eds. B.Ebenman and L. Persson), pp. 203–18. Springer, Berlin.

Persson, L. and Greenberg, L. (1990). Juvenile competitive bottlenecks: the perch (*Perca fluviatilis*)–roach (*Rutilus rutilus*) interaction. *Ecology*, **71**, 44–56.

Persson, L., Andersson, G., Hamrin, S. F., and Johansson, L. (1988). Predator regulation and primary productivity along the productivity gradient of temperate lake ecosystems. In *Complex interactions in lake communities*, (ed. S. R. Carpenter), pp. 45–65. Springer, New York.

Petchey, O., McPhearson, P. T., Casey, T. M., and Morin, P. J. (1999). Environmental warming alters food-web structure and ecosystem function. *Nature*, **402**, 69–72.

Petranka, J. W., Kats, L. B., and Sih, A. (1987). Predator–prey interactions among fish and larval amphibians: use of chemical cues to detect predators. *Animal Behaviour*, **35**, 420–5.

Pettersson, L. B. and Brönmark, C. (1997). Density-dependent costs of an inducible morphological defense in crucian carp. *Ecology*, **78**, 1805–15.

Pimentel, D., Lach, L., Zuniga, R., and Morrison, D. (2000). Environmental and economic costs of nonindigenous species in the United States. *Bioscience*, **50**, 53–65.

Polis, G. A. and Strong, D. R. (1997). Food web complexity and community dynamics. *American Naturalist*, **147**, 813–46.

Porter, K. (1973). Selective grazing and differential digestion of algae by zooplankton. *Nature*, **244**, 179–80.

Raffaelli, D., van der Putten, W. H., Persson, L., Wardle, D. A., Petchey, O. L., Koricheva, J., van der Heijden, M., Mikola, J., and Kennedy, T. (2002). Multi-trophic dynamics and ecosystem processes. In *Biodiversity and ecosystem functioning*, (eds. M. Loreau, S. Naeem and P. Inchausti), pp.147–54. Oxford University Press, Oxford.

Rahel, F. J. (2002). Homogenisation of freshwater faunas. *Annual Review of Ecology and Systematics*, **33**, 291–315.

Reichhart, T. (2000). Will souped up salmon sink or swim? *Nature*, **406**, 10–2.

Reynolds, C. S. (1984). *The ecology of freshwater phytoplankton*. Cambridge University Press.

Ricciardi, A. (2001). Facilitative interactions among aquatic invaders: is an 'invasional meltdown' occurring in the Great Lakes? *Canadian Journal of Fisheries and Aquatic Sciences*, **58**, 2513–25.

Ricciardi, A. and Rasmussen, J. B. (1999). Extinction rates of North American freshwater fauna. *Conservation Biology*, **13**, 1220–2.

Richter, I. D., Braun, D. P., Mendelson, M. A., and Master, L. L. (1997). Threats to imperiled freshwater fauna. *Conservation Biology*, **11**, 1081–93.

Rogers, D. and Randolph, S. (2000). The global spread of malaria in a future, warmer world. *Science*, **289**, 1763–6.

Romanovsky, Y. E. and Feniova, I. Y. (1985). Competition among cladocera: effect of different levels of food supply. *Oikos*, **44**, 243–52.

Romare, P. and Hansson, L-A. (2003). A behavioral cascade: top-predator induced behavioral shifts in planktivorous fish and zooplankton. *Limnology and Oceanography*, **48**, 1956–64.

Room, P. M. (1990). Ecology of a simple plant–herbivore system: biological control of *Salvinia*. *Trends of Ecology and Evolution*, **5**, 74–9.

Rothhaupt, K–O. (1996). Laboratory experiments with a mixotrophic chrysophyte and obligately phagotrophic and photographic competitors. *Ecology*, **77**, 716–24.

Rowell, K. and Blinn, D. W. (2003). Herbivory on a chemically defended plant as a predation deterrent in *Hyalella azteca*. *Freshwater Biology*, **48**, 247–54.

Rundel, P., Ehleringer, J., and Nagy, K. (eds.) (1988). *Stable isotopes in ecological research*. Springer, Berlin.

Sala, O. E., Chapin F. S., III, Armesto, J. J., Berlow, E., Bloomfield, J., Dirzo, R., Huber-Sanwald, E., Huenneke, L. F., Jackson, R. B., Kinzig, A., Leemans, R., Lodge, D. M., Mooney, H. A., Oesterheld, M., Poff, N. L., Sykes, M. T., Walker, B. H., Walker, M., and Wall, D. H. (2000). Global biodiversity scenarios for the year 2100. *Science*, **287**, 1770–4.

Sand-Jensen, K. and Borum, J. (1984). Epiphyte shading and its effect on photosynthesis and diel metabolism of *Lobelia dortmanna* L. during the spring bloom in a Danish lake. *Aquatic Botany*, **20**, 109–19.

Savino, J. F. and Stein, R. A. (1989). Behavioural interactions between fish predators and their prey: effects of plant density. *Animal Behaviour*, **37**, 311–21.

Scheffer, M. (1990). Multiplicity of stable states in freshwater systems. *Hydrobiologia*, **200/201**, 475–86.

Scheffer, M. and Carpenter, S. R. (2003). Catastrophic regime shifts in ecosystems: linking theory to observation. *Trends in Ecology and Evolution*, **18**, 648–56.

Schindler, D. E., Carpenter, S. R., Cole, J. J., Kitchell, J. F., and Pace, M. L. (1997). Influence of food web structure on carbon exchange between lakes and the atmosphere. *Science*, **277**, 248–51.

Schindler, D. W. (1974). Eutrophication and recovery in experimental lakes: implications for lake management. *Science*, **184**, 897–9.

Schindler, D. W., Curtis, P. J., Parker, B. R., and Stainton, B. R. (1996). Consequences of climate warming and lake acidification for UV-B penetration in North American boreal lakes. *Nature*, **379**, 705–8.

Scrimshaw, S. and Kerfoot, W. C. (1987). Chemical defenses of freshwater organisms: beetles and bugs. In Predation: *direct and indirect impacts on aquatic communities*, (eds. W. C. Kerfoot and A. Sih), pp. 240–62. University Press of New England, Hanover, NH.

Seehausen, O., van Alphen, J. J. M., and Witte, F. (1997). Cichlid fish diversity threatened by eutrophication that curbs sexual selection. *Science*, **277**, 1808–11.

Semlitsch, R. D., Scott, D. E., and Pechmann, J. H. K. (1988). Time and size at metamorphosis related to adult fitness in *Ambystoma talpoideum*. *Ecology*, **69**, 184–92.

Shapiro, J., Lamarra, V., and Lynch, M. (1975). Biomanipulation: an ecosystem approach to lake restoration. In *Proceedings of a symposium on water quality management through biological control*, (eds. P. L. Brezonik and J. L. Fox), pp. 85–96. University of Florida, Gainesville, FL.

Shurin, J. B. (2000). Dispersal limitation, invasion resistance, and the structure of pond zooplankton communities. *Ecology*, **81**, 3074–86.

Sih, A. (1980). Optimal behavior: can foragers balance two conflicting demands. *Science*, **210**, 1041–3.

Sih, A., Englund, G., and Wooster, D. (1998). Emergent impacts of multiple predators on prey. *Trends in Ecology and Evolution*, **13**, 350–5.

Simberloff, D. (2001). Introduced species, effects and distribution. *Encyclopedia of Biodiversity*, **3**, 517–29.

Simberloff, D. and Von Holle, B. (1999). Positive interactions of nonindigenous species: invasional meltdown? *Biological Invasions*, **1**, 21–32.

Slusarczyk, M. (1995). Predator-induced diapause in *Daphnia*. *Ecology*, **76**, 1008–13.

Sommer, U. (1989). *Plankton ecology: succession in plankton communities*. Springer, Berlin.

Sommer, U., Gliwicz, Z. M., Lampert, W., and Duncan, A. (1986). The PEG-model of seasonal succession of planktonic events in freshwaters. *Archiv für Hydrobiologie*, **106**, 433–71.

Soszka, G. J. (1975). The invertebrates on submerged macrophytes in three Masurian lakes. *Ekologia Polska*, **23**, 371–91.

Southwood, T. R. E. (1988). Tactics, strategies and templets. *Oikos*, **52**, 3–18.

Stabell, T., Andersen, T., and Klaveness, D. (2002). Ecological significance of endosymbionts in a mixotrophic ciliate–an experimental test of a simple model of growth coordination between host and symbiont. *Journal of Plankton Research*, **24**, 889–99.

Ståhl-Delbanco, A. and Hansson, L.-A. (2002). Effects of bioturbation from benthic invertebrates on recruitment of algal resting stages from the sediment. *Limnology and Oceanography*, **47**, 1836–43.

Stein, R. A. and Magnuson, J. J. (1976). Behavioral response of crayfish to a fish predator. *Ecology*, **57**, 751–61.

Stemberger, R. S. and Gilbert, J. J. (1987). Defenses of planktonic rotifers against predators. In *Predation. Direct and indirect impacts in aquatic communities* (eds. W. C. Kerfoot and A. Sih), pp. 227–39. University Press of New England, Hanover, NH.

Stenson, J. A. E., Svensson, J.-E., and Cronberg, G. (1993). Changes and interactions in the pelagic community in acidified lakes in Sweden. *Ambio*, **22**, 277–82.

Stephens, D. W. and Krebs, J. R. (1986). *Foraging theory*. Princeton University Press, Princeton NJ.

Sterner, R. W. and Elser, J. J. (2002). *Ecological stoichiometry. The biology of elements from molecules to the biosphere*. Princeton University Press, Princeton, NJ.

Stewart, T. W., Miner, J. G., and Lowe, R. L. (1998). Quantifying mechanisms for zebra mussel effects on benthic macroinvertebrates: organic matter production and shell-generated habitat. *Journal of North American Benthological Society*, **17**, 81–94.

Stich, H-B. and Lampert, W. (1981). Predator evasion as an explanation of diurnal vertical migration by zooplankton. *Nature*, **293**, 396–8.

Stoddard, J. L., Jeffries, D. S., Lükewille, A., Clair, T. A, Dillon, P. J., Driscoll, C. T., Forsius, M., Johannessen, M., Kahl, J. S., Kellogg, J. H., Kemp, A., Mannio, J., Monteith, D. T., Murdoch, P. S., Patrick, S., Rebsdorf, A., Skjelkvåle, B. L., Stainton, M. P., Traaen, T., van Dam, H., Webster, K. E., Wieting, J., and

Wilander, A. (1999). Regional trends in aquatic recovery from acidification in North America and Europe. *Nature*, **401**, 575–8.

Straile, D. (2002). North Atlantic Oscillation synchronizes food-web interactions in central European lakes. *Proceedings of the Royal Society of London, Series B*, **2669**, 391–5.

Strayer, D. (2001). Endangered freshwater invertebrates. *Encyclopedia of Biodiversity*, **2**, 425–39.

Streams, F. A. (1987). Within-habitat spatial separation of two *Notonecta* species: interactive vs. noninteractive resource partitioning. *Ecology*, **68**, 935–45.

Strong, D. (1992). Are trophic cascades all wet? Differentiation and donor control in speciose ecosystems. *Ecology*, **73**, 747–54.

Suttle, C. A. (1994). The significance of viruses to mortality in aquatic microbial communities. *Microbial Ecology*, **28**, 237–243.

Szöllosi-Nagy, A., Najlis, A., and Björklund, G. (1998). Assessing the world's freshwater resources. *Nature and Resources*, **34**, 8–18.

Thingstad, T. F. (2000). Elements of a theory for the mechanisms controlling abundance, diversity, and biogeochemical role of lytic bacterial viruses in aquatic systems. *Limnology and Oceanography*, **45**, 1320–8.

Tilman, D. (1980). Resources: a graphical-mechanistic approach to competition and predation. *American Naturalist*, **116**, 362–93.

Tilman, D. (1982). Resource competition and community structure. *Monographs in Population Biology*. Princeton University Press, Princeton NJ.

Tilman, D., Kiesling, R., Sterner, R., Kilham, S. S., and Johnson, F. A. (1986). Green, blue-green and diatom algae: taxonomic differences in competitive ability for phosphorus, silicon and nitrogen. *Archiv für Hydrobiologie*, **106**, 473–85.

Tittel, J., Bissinger, V., Zippel, B., Gaedke, U., Bell, E., Lorke, A., and Kamjunke, N. (2003). Mixotrophs combine resource use to outcompete specialists: implications for aquatic food webs. *Proceedings of the National Academy of Sciences, USA*, **28**, 12776–81.

Tonn, W. M., Magnuson, J. J., Rask, M., and Toivonen, J. (1990). Intercontinental comparison of small-lake fish assemblages: the balance between local and regional processes. *American Naturalist*, **136**, 345–75.

Tonn, W. M., Paszkowski, C. A., and Holopainen, I. J. (1992). Piscivory and recruitment: mechanisms structuring prey populations in small lakes. *Ecology*, **73**, 951–8.

Tonn, W. M., Holopainen, I. J., and Paszkowski, C. A. (1994). Density-dependent effects and the regulation of crucian carp populations in single-species ponds. *Ecology*, **75**, 824–34.

Tranvik, L. J. (1988). Availability of dissolved organic carbon for planktonic bacteria in oligotrophic lakes of differing humic content. *Microbial Ecology*, **16**, 311–22.

Travis, J. (1983). Variation in growth and survival of *Hyla gratiosa* larvae in experimental enclosures. *Copeia*, **1983**, 232–7.

Turner, A. M. (1996). Freshwater snails alter habitat use in response to predation. *Animal Behaviour*, **51**, 747–56.

Turner, A. M. (1997). Contrasting short-term and long-term effects of predation risk on consumer habitat use and resources. *Behavioural Ecology*, **8**, 120–5.

Turpin, D. H. (1991). Physiological mechanisms in phytoplankton resource competition. In *Growth and reproductive strategies of freshwater phytoplankton*, (ed. C. D. Sandgren), pp. 316–68. Cambridge University Press.

Vadeboncoeur, Y., Vander Zanden, M. J., and Lodge, D. M. (2002). Putting the lake back together: reintegrating benthic pathways into lake food web models. *Bioscience*, **52**, 44–54.

Vander Zanden, M. J. and Vadenboncoeur, Y. (2002). Fishes as integrators of benthic and pelagic food webs in lakes. *Ecology*, **83**, 2151–61.

Vander Zanden, M. J., Shuter, B. J., Lester, N., and Rasmussen, J. B. (1999). Patterns of food chain length in lakes: a stable isotope study. *American Naturalist*, **154**, 406–16.

Vanni, M. (1996). Nutrient transport and recycling by consumers in lake food webs: implications for algal communities. In *Food webs. Integration of patterns and dynamics*, (eds. G. A. Polis and K. O. Winemiller), pp. 81–95. Chapman & Hall, New York.

Vanni, M. J., Luecke, C., Kitchell, J. F., Allen, Y., Temte, J., and Magnusson, J. J. (1990). Effects on lower trophic levels of massive fish morality. *Nature*, **344**, 333–5.

Vershuren, D., Johnson, T. C., Kling, H. J., Edgington, D. N., Leavitt, P. R., Brown, E. T., Talbot, M. R., and Hecky, R. E. (2002). History and timing of human impact on Lake Victoria, east Africa. *Proceedings of the Royal Society London, Series B*, **269**, 289–94.

Vollenweider, R. A. (1968). *Scientific fundamentals of the eutrophication of lakes and flowing waters, with particular reference to nitrogen and phosphorus as factors in eutrophication*. DAS/CSI/68.27. Organization for Economic Cooperation and Development (OECD), Paris.

Voltaire, F. M. A. (1759). *Candide, où la optimisme*. Cramer, Geneva.

Wahl, D. H. and Stein, R. A. (1988). Selective predation by three esocids: the role of prey behavior and morphology. *Transactions of the American Fisheries Society*, **117**, 142–51.

Wainwright, P. C., Osenberg, C. W., and Mittelbach, G. G. (1991). Trophic polymorphism in the pumpkinseed sunfish (*Lepomis gibbosus* Linnaeus): effects of environment on ontogeny. *Functional Ecology*, **5**, 40–55.

Walker, B. H. (1992). Biodiversity and ecological redundancy. *Conservation Biology*, **6**, 18–23.

Wania, F. and Mackay, D. (1993). Global fractionation and cold condensation of low volatility organochlorine compounds in polar regions. *Ambio*, **22**, 10–8.

Weatherly, N. S. (1988). Liming to mitigate acidification in freshwater ecosystems: a review of the biological consequences. *Water, Air and Soil Pollution*, **39**, 421–37.

Webb, P. W. (1984). Body form, locomotion and foraging in aquatic vertebrates. *American Zoologist*, **24**, 107–20.

Wedekind, C. and Milinski, M. (1996). Do three-spined sticklebacks avoid consuming copepods, the first intermediate host of *Schistocephalus solidus*? An experimental analysis of behavioural resistance. *Parasitology*, **112**, 371–83.

Weisner, S. E. B. (1987). The relation between wave exposure and distribution of emergent vegetation in a eutrophic lake. *Freshwater Biology*, **18**, 537–44.

Weisner, S. E. B. (1993). Long term competitive displacement of *Typha latifolia* by *Typha angustifolia* in a euthropic lake. *Oecologia*, **94**, 451–6.

Werner, E. E. (1988). Size, scaling, and the evolution of complex life cycles. In *Size-structured populations. Ecology and evolution*, (eds. B. Ebenman and L. Persson), pp. 60–81. Springer, Berlin.

Werner, E. E. and Hall, D. J. (1974). Optimal foraging and the size selection of prey by the bluegill sunfish (*Lepomis macrochirus*). *Ecology*, **55**, 1042–52.

Werner, E. E., Gilliam, J. F., Hall, D. J., and Mittelbach, G. G. (1983). An experimental test of the effects of predation risk on habitat use in fish. *Ecology*, **64**, 1540–8.

Wetzel, R. G. and Likens, G. E. (1991). *Limnological analyses.* Springer-Verlag, New York.

Wiggins, G. B., Mackay, R. J., and Smith, I. M. (1980). Evolutionary and ecological strategies of animals in annual temporary pools. *Archiv für Hydrobiologie*, Supplement, **58**, 97–206.

Wilbur, H. M. (1976). Density-dependent aspects of metamorphosis in *Ambystoma* and *Rana sylvatica. Ecology*, **57**, 1289–96.

Wilbur, H. M. (1980). Complex life cycles. *Annual Review of Ecology and Systematics*, **11**, 67–93.

Wilbur, H. M. (1984). Complex life cycles and community organization in amphibians. In *A new ecology: novel approaches to interactive systems* (eds. S. W. Price, C. N. Sloboschikoff, and W. S. Gaud), pp. 195–233. Wiley, New York.

Williamson, C. E. (1995). What role does UV-B radiation play in freshwater ecosystems? *Limnology and Oceanography*, **40**, 386–392.

Williamson, C. E., Stemberger, R. S., Morris, D. P., Frost, T. M., and Paulsen, S. G. (1996). Ultraviolet radiation in North American lakes: attenuation estimates from DOC measurements and implications for plankton communities. *Limnology and Oceanography*, **41**, 1024–34.

Wimberger, P. H. (1992). Plasticity of fish body shape. The effects of diet, development, family and age in two species of *Geophagus* (Pisces: Cichlidae). *Biological Journal of the Linnaean Society*, **45**, 197–218.

Wommack, K. E. and Colwell, R. R. (2000). Virioplankton: Viruses in aquatic ecosystems. *Microbiology and Molecular Biology Reviews*, **64**, 69–114.

Xenopoulos, M. A. and Bird, D. F. (1997). Effects of acute exposure to hydrogen peroxide on the production of phytoplankton and bacterioplankton in a mesohumic lake. *Photochemistry and Photobiology*, **66**, 471–8.

Yentsch, C. (1974). Some aspects of the environmental physiology of marine phytoplankton: a second look. *Oceanography Marine Biology Annual Reviews*, **12**, 41–75.

Zaret, T. M. (1972). Predators, invisible prey, and the nature of polymorphism in the cladocera (class Crustacea). *Limnology and Oceanography*, **17**, 171–84.

Zedler, J. B. (2003). Wetlands at your service: reducing impacts of agriculture at the watershed scale. *Frontiers in Ecology and the Environment*, **1**, 65–72.

Further Reading

This list of selected literature is for those wanting more information on a specific subject.

Limnological methods

Barnard, C., Gilbert, F., and McGregor, P. (1993). *Asking questions in biology— design, analysis and presentation in practical work*. Longman, Harlow, Essex, UK.

Hutchinson, G. E. (1975). *A treatise on limnology*. Vol. III. *Limnological methods*. Wiley, New York.

Rosenberg, D. M. and Resh, V. H. (eds.) (1993). *Freshwater biomonitoring and benthic macroinvertebrates*. Chapman & Hall, New York.

Schreck, C. B. and Moyle, P. B. (eds.) (1990). *Methods for fish biology*. American Fisheries Society, Bethesda, MD.

Underwood, A. J. (1997). *Experiments in ecology*. Cambridge University Press.

Wetzel, R. G. and Likens, G. E. (1991). *Limnological analyses*. Springer-Verlag, New York.

The abiotic frame

Håkansson, L. and Jansson, M. (1983). *Principles of lake sedimentology*. Springer, Berlin.

Hutchinson, G. E. (1957). *A treatise on limnology*. Vol. I. *Geography, physics and chemistry*. Wiley, New York.

Kirk, J. T. O. (1983). *Light and photosynthesis in aquatic ecosystems*. Cambridge University Press.

Sterner, R. W. and Elser, J. J. (2002). *Ecological stoichiometry, the biology of elements from molecules to the biosphere*. Princeton Univeristy Press, Princeton, NJ.

Williams, D. D. (1987). *The ecology of temporary waters*. Croom Helm, Beckenham, UK.

Wetzel, R. G. (2001). Limnology. 3rd ed. Academic Press, SanDiego, Ca.

The organisms

Canter-Lund, H. and Lund, J. W. G. (1995). *Freshwater algae—their microscopic world explored*. Biopress, Bristol, UK.

Dusenbery, D. B. (1996). *Life at a small scale—the behavior of microbes*. Scientific American Library, New York.

Hutchinson, G. E. (1967). *A treatise on limnology*. Vol. II. *Introduction to lake biology and the limnoplankton*. Wiley, New York.

Hutchinson, G. E. (1993). *A treatise on limnology*. Vol. IV. *The zoobenthos*. Wiley, New York.

Maitland, P. S. and Campbell, R. N. (1992). *Freshwater fishes*. HarperCollins, London.

Nilsson, A. (1996). *Aquatic insects of north Europe—a taxonomic handbook*, Vols. I and II. Apollo Books, Stenstrup, Denmark.

Owen, M. and Black, J. M. (1990). *Waterfowl ecology*. Blackie and Son, Ltd, Glasgow, UK.

Patterson, D. J. and Hedley, S. (1992). *Free-living freshwater protozoa—a colour guide*. Wolfe, Aylesbury, UK.

Reynolds, C. S. (1984). *The ecology of freshwater phytoplankton*. Cambridge University Press.

Sandgren, C. D. (1991). *Growth and reproductive strategies of freshwater phytoplankton*. Cambridge University Press.

Stebbins, R. C. and Cohen, N. W. (1995). *A natural history of amphibians*. Princeton University Press, Princeton, NJ.

Thorp, J. H. and Covich, A. P. (eds.) (1991). *Ecology and classification of North American freshwater invertebrates*. Academic Press, San Diego, CA.

Wootton, R. J. (1990). *Ecology of teleost fishes*. Chapman & Hall, London.

The Freshwater Biological Association has produced a series of identification keys to a whole range of freshwater organisms. These can be obtained from the Freshwater Biological Association, The Ferry House, Far Sawrey, Ambleside, Cumbria, LA22 OLP, UK.

Biotics

Harris, G. P. (1986). *Phytoplankton ecology—structure, function and fluctuation*. Chapman & Hall, New York.

Keddy, P. A. (1989). *Competition*. Chapman & Hall, New York.

Kerfoot, W. C. and Sih, A. (eds.) (1987). *Predation. Direct and indirect impacts on aquatic communities*. University Press of New England, Hanover, NH.

Laybourn-Parry, J. (1992). *Protozoan plankton ecology*. Chapman & Hall, London.

Stevenson, R. J., Bothwell, M. L., and Lowe, R. L. (eds.) (1996). *Algal ecology— freshwater benthic ecosystems*. Academic Press, San Diego, CA.

Tilman, D. (1982). *Resource competition and community structure*. Monographs in population biology, No. 17. Princeton University Press, Princeton, NJ.

Food web interactions

Carpenter, S. R. (ed.) (1987). *Complex interactions in lake communities*. Springer, New York.

Carpenter, S. R. and Kitchell, J. F. (1993). *The trophic cascade in lakes*. Cambridge University Press.

Polis, G. A. and Winemiller, K. O. (eds.) (1996). *Food webs. Integration of patterns and dynamics*. Chapman & Hall, New York.

Scheffer, M. (1998). *Ecology of shallow lakes*. Chapman & Hall, London.

The environment and conservation

Goldschmidt, T. (1996). *Darwin's dreampond. Drama in Lake Victoria*. MIT Press, Cambridge, MA.

Kitchell, J. F. (ed.) (1992). *Food web management—a case study of Lake Mendota*. Springer, New York.

Loreau, M., Naeem, S., and Inchausti, P. (2002). *Biodiversity and ecosystem functioning*. Oxford University Press, Oxford.

Glossary

This section includes most of the words appearing in **bold face** in the text, as well as several others which may be useful when trying to penetrate the jargon of aquatic ecology.

Abiotic factors physical and chemical factors

Adaptation a genetically determined trait that makes an organism more successful in its environment

Alkalinity the capacity of a specific water to neutralize acid, which is usually a result of the amount of carbonate, bicarbonate, and hydroxide content

Allelopathy the release of chemical substances by plants that negatively affect other plants

Allochthonous material organic material produced outside the lake

Amensalism interaction between two organisms where one is negatively affected, whereas the other is not affected in any way

Anion a negatively charged ion

Anterior front

Aphotic zone the depth beyond which less than 1% of the surface light reaches; light availability is insufficient for photosynthesis

Asymmetric competition when one competitor is more competitive than the other

ATP (adenosine triphospate) the main energy carrier in cell metabolism

Aufwuchs German word for microflora (algae, bacteria, fungi) growing on solid substrates; synonymous to **Periphyton**

Autochthonous material organic matter produced within the lake

Autotrophic organism an organism that fixes inorganic carbon using light as energy source

Biomass total mass of living organisms per unit of surface or volume

Biodiversity the extent of genetic, taxonomic, and ecological variability over all spatial and temporal scales

Biotic interactions interactions among living organisms (e.g. competition, predation, symbiosis)

Bioturbation when an organism physically mix, for example, the sediment of a lake

CAM (crassualacean acid metabolism) a subsidary mechanism to photosynthesis, allowing night-time uptake of carbon dioxide that can be stored for daytime use in photosynthesis

Catchment area the region around the lake or pond that drains the rain to the lake

Cation a positively charged ion

Clearwater phase a period, often during spring, when many lakes become clear due to low algal turbidity

Compensatory depth the depth where photosynthesis equals respiration

Detritus dead organic matter

Detrivore an organism that feeds on detritus

Diapause a dormancy period in the life cycle of an organism

DIC dissolved inorganic carbon

Dimictic lake a lake that undergoes complete mixing twice a year

DOC dissolved organic carbon

Dorsal on the back

Drainage area *see* **Catchment area**

Dystrophic lake a lake in which production is mainly based on allochthonous material (from the Greek: *dystrophus* = mal nutritious)

Ecosystem all organisms and their environment in a specific location, such as a lake or a pond

Ectoparasite a parasite that resides on the outside of its host

Endoparasite a parasite that resides inside its host

Ephippium the thick-walled chamber for fertilized resting eggs of many cladocerans

Epi- the prefix means 'on'

Epibiont an organism that lives attached to another organism

Epilimnion the stratum of warm, well-mixed water above the thermocline

Epipelic growing on sediments

Epiphytic growing on plants

Episammic growing on sand

Eukaryote an organism with a nucleus

Eurytherm an organism that tolerates a wide temperature range

Eutrophic lake nutrient-rich, high productivity lake where production is mainly based on internally produced carbon; that is, photosynthesis (from the Greek: *eutrophus* = high nutritious)

Exploitation competition competition between organisms mediated through the reduction of common, limiting resources

External loading input of nutrients to a lake from its surroundings

Fitness a measure of the relative ability of a genotype to transfer genes to the next generation

Food chain the idealized linear relationship of energy flow from primary producers, through herbivores, to predators

Functional group organisms that feed in a similar way (e.g. filter-feeders, grazers, or predators)

Functional response the relation between individual consumption rate and resource density

Fundamental niche the range of environmental conditions within which a population can live and reproduce in the absence of other organisms

Gross photosynthesis the sum of net photosynthesis and respiration

Guild group of populations exploiting the same resources in a similar way

Herbivore organism that feeds on primary producers

Heterocyte a specialized cell where nitrogen fixation takes place in some cyanobacteria

Heterophylly when individual plants have leaves that differ in morphology depending on whether they are submersed or emergent

Heterotrophic an organism that uses organic carbon sources (other organisms or detritus) to build-up biomass

Homeostatic consumer a consumer that keeps its chemical composition constant irrespective of the chemistry of its food

Humic substances large, refractory molecules mainly formed as a result of decomposition of organic material, giving 'humic lakes' their brownish colour

Hypolimnion the stratum of water below the thermocline

Integumental respiration uptake of oxygen directly across the body surface

Interference competition competition where access to a limited resource is reduced by direct interference between competitors

Internal loading input of nutrients to the lake or pond from its own sediment

Interspecific competition competition between individuals of different species

Intraspecific competition competition between individuals of the same species

Isocline a depth isocline is a line connecting sites with similar depth

Kairomone a chemical substance released by one organism that elicits a response (behavioural or morphological) in another organism

Keystone species a species that has a disproportionately strong effect on other species and therefore has a significant role in community organization

Kettle lake a lake formed by an ice block left behind by a retarding glacier. When the ice block eventually melted, a small lake was formed. Also called pothole lake

Lateral side

Lateral line a sensory line along the side of fishes, used to detect vibrations

Limnology the study of freshwater communities and their interactions with physical, chemical, and biotic environmental variables

Littoral zone the shallow, nearshore part of a lake or pond characterized by light penetration to the bottom

Lysis cell rupture, for example, due to virus infection

Macrophytes higher aquatic plants

Mastax muscular 'stomach' of rotifers

Metalimnion the stratum between the epilimnion and hypolimnion exhibiting a marked thermal change; synonymous to **Thermocline**

Metamorphosis marked change in morphology during ontogeny

Metazoa multicellular eukaryotes

Microbe unicellular organism

Mixotrophic organism an organism that is able to use both photosynthesis (autotrophy) as well as assimilating organic compounds (heterotrophy) for its energy support

Monomictic lake a lake that undergoes complete mixing once a year

Mutualism interaction between two populations where both benefit from the association

Net photosynthesis gross photosynthesis minus respiration

Nitrification bacterial transformation of ammonium to nitrate

Oligotrophic lake nutrient-poor, low productivity lake where production is mainly based on internally produced carbon (from the Greek: *oligotrophus* = low nutritious)

Omnivore an organism that feeds on organisms from different trophic levels (e.g. plants and herbivores)

Ontogeny the development of an individual organism over its life cycle

Parasite an organism feeding on another without immediately killing it; the parasite is generally much smaller than the host

Parthenogenesis the development of eggs into fully grown individuals without fertilization

Pelagic zone the open water part of a lake or pond

Periphyton microorganisms living on surfaces

Phagotrophic feeding on other organisms to obtain carbon (cf. phototrophic)

Photic zone where the light intensity is above 1% of surface light; where light availability is high enough to allow photosynthesis

Phototrophic an organism using photosynthesis as carbon source (cf. phagotrophic)

Photosynthesis the process in plants where large organic molecules are produced from inorganic carbon using light as the energy source

Phototactic organism organism that moves towards light

Phytoplankton primary producers that float freely in the water

Piscivore an organism that eats fish

Planktivore an organism that eats zooplankton

Plankton an organism that is suspended in the water

Poikilothermic organism an organism that has a similar temperature as the environment

Polymictic a lake that undergoes complete mixing more than twice a year

Pond a small, shallow waterbody where temperature changes are more important for mixing than wind

Posterior behind

Pothole lake a lake formed by an ice block left behind by a retarding glacier. When the ice block eventually melted, a small lake was formed. Also called **Kettle lake**

Predator an organism that feeds on animals

Primary producer organism that transforms light energy into chemical energy by photosynthesis

Profundal zone the sediment zone at depths beyond where primary producers can live

Prokaryote an organism without a nucleus or other discrete cellular organelles

Protozoa unicellular eukaryotes distinguished from algae by obtaining energy and nutrients by heterotrophy instead of by photosynthesis. Protozoa includes amoeboid, flagellated, and ciliated organisms capable of heterotrophic nutrition.

Realized niche the range of environmental conditions occupied by a population in the presence of other organisms

Resource factor that results in increased growth as availability is increased

Redfield ratio the atomic ratio of the elements carbon, nitrogen, and phosphorus $(C : N : P)$ of $106 : 16 : 1$

Redox potential an estimate of the total intensity of oxidizing or reducing chemical reactions and mirrors the relative availability of electrons in a solution

Saphrophyte plant obtaining food from dead or decaying organisms

Seiche wind-induced tilting of a lake's water level and thermocline. When the wind speed is reduced, the water level 'tilts back', causing a standing wave

Secchi depth an estimate of the light penetration through the water made by lowering a circular plate until the plate is no longer visible. This depth is the Secchi depth

Spicula needle-shaped silica formation in freshwater sponges

Stenotherm an organism that has a limited temperature range

Succession temporal change in community composition

Symbiosis a mutually beneficial interaction between two different species; synonymous with **Mutualism**

Thermocline the stratum between the epilimnion and hypolimnion exhibiting a marked thermal change; synonymous with **Metalimnion**

Trophic level functional classification of organisms in a food web (food chain) with respect to their feeding relationships. Traditionally, the first trophic level is primary producers, the second herbivores, and the third small predators (e.g. planktivorous fish), and the fourth predators feeding on small predators (e.g. piscivorous fish)

Trophy nutrient status of a lake or pond

Vector an organism that transmits a pathogen (e.g. virus, bacteria, or protozoan), to another organism

Ventral below

Zooplankton floating and drifting animal life

Index

Page numbers in *italics* refer to material in Figures.